A DINGO ATE MY MATH BOOK

Mathematics from Down Under

A DINGO ATE MY MATH BOOK

Mathematics from Down Under

Burkard Polster
Marty Ross

AMERICAN MATHEMATICAL SOCIETY
Providence, Rhode Island

2010 *Mathematics Subject Classification*. Primary 00A08, 00A09, 00A67, 97A80.

For additional information and updates on this book, visit
www.ams.org/bookpages/mbk-106

Library of Congress Cataloging-in-Publication Data

Names: Polster, Burkard. | Ross, Marty, 1959-
Title: A dingo ate my math book : mathematics from Down Under / Burkard Polster, Marty Ross.
Description: Providence, Rhode Island : American Mathematical Society, [2017]
Identifiers: LCCN 2017014890 | ISBN 9781470435219 (alk. paper)
Subjects: LCSH: Mathematics–Australia–Popular works. | AMS: General – General and miscellaneous specific topics – Recreational mathematics. msc | General – General and miscellaneous specific topics – Popularization of mathematics. msc | General – General and miscellaneous specific topics – Mathematics and architecture. msc | Mathematics education – General, mathematics and education – Popularization of mathematics. msc
Classification: LCC QA93 .P65 2017 | DDC 510–dc23 LC record available at https://lccn.loc.gov/2017014890

∞ The paper used in this book is acid-free and falls within the guidelines
established to ensure permanence and durability.
Visit the AMS home page at http://www.ams.org/

10 9 8 7 6 5 4 3 2 1 22 21 20 19 18 17

This book is dedicated to Simon Pryor, spruiker extraordinaire

Contents

Preface

From 2007 to 2014 we produced the weekly "Maths Masters" column for Melbourne's *Age* newspaper. (Australians follow the English in studying "maths" rather than "math".) This book comprises 1,000,000 of those columns, selected from the 11,111,111 columns we wrote in total.

The mission of our column, and of the book you are now holding, was to present ingenious, unusual and beautiful nuggets of mathematics in as clear and as entertaining a manner as possible. Since the column appeared in a regular newspaper, we had to assume that our readers had little mathematics background apart from, perhaps, a few remnants left over from school days. As such, most columns employed a little arithmetic or geometry or algebra, but little else. Still, we were often ambitious, exploring genuinely deep ideas and—bloody hell!—proving things. Of course this often required simplification and shortcuts, but we always endeavored to be honest, to never cheat our readers.

What did we write about? Whatever took our fancy: some columns were our versions of familiar standards of math popularisation; others were inspired by news items, either mathematical breakthroughs or pseudo-mathematical nonsense; others were the product of our random pondering, as we pursued our mathematician activities.

Many "Maths Masters" columns had a distinctly Australian flavor, and it is a selection of those columns that we have chosen for this book. Some of this Aussieness is gratuitous, the framing of general ideas in a (hopefully) entertaining scenario. Other columns are much more genuinely Australian, focusing upon, for example, Australians' love of sports and, alas, gambling, or upon Melbourne's iconic, mathematically inspired architecture.

Our column changed significantly over the years. It began in print, with a limit of about 400 words and little room for graphics: just enough space to say hello, introduce an idea, and wave goodbye again. In 2009 the column moved online, allowing us much more freedom to explore ideas in depth.

The column also changed in another significant way. In 2009 Australia's education authorities began the design and implementation of a new, national curriculum. Dominated by education "experts", and with the almost total exclusion of mathematicians, the implementation was farcical, and the mathematics curriculum that emerged is simply a mess: strong on calculators and foolish pseudo-relevance, and very, very short on genuine mathematics, on beauty and reasoning and proof. This state of affairs inspired a number of grumpy columns criticizing the mathematics curriculum and Australia's math education in general.

The columns as reproduced here are pretty much as they first appeared in *The Age*, except for updating, and other minor corrections and adjustments. We've included explanatory footnotes when it seemed helpful or necessary, and each column ends with a (possibly tricky) puzzle or two; solutions to the puzzles are provided in the appendix.

This book is dedicated to Simon Pryor, then CEO of the Mathematical Association of Victoria, for his incredible—sometimes bemused—support over the years. For over a decade Simon was our cheerleader, roadie, agent and God-knows-what-else, and it was Simon who shouted our names when *The Age* went hunting for writers of a new mathematics column. Many of the opportunities that we have enjoyed would never have eventuated if not for Simon's inexhaustible efforts.

We would like to thank the following people and organisations for their very kind permission to reproduce photographs in the book: Bruce Bilney for photos of his tessellations (Chapter 1); the University of California, Los Angeles, for the photo of Professor Terry Tao (Chapter 13); Feliks Zemdegs for the photo of himself speedcubing (Chapter 14); Rayda Deakin for the photo of her late husband Michael Deakin (Chapter 15); and architects McBride Charles Ryan for the photos of the Klein Bottle house (Chapter 22).

We'd also like to thank the wonderful Ken Merrigan, our original editor at *The Age*, who guided us with patient good humor through our first clumsy years. Thanks also to Ben Haywood at *The Age*, who somehow coped with the naively ambitious typography and graphics that we threw at him on a weekly basis. Thank you to our editor Ina Mette at the AMS and all her colleagues who helped with making this book a reality. And, a huge thanks to all our readers, whose patronage ensured the long and happy life of Australia's only mathematics column.

Finally, our thanks and apologies to Anu and Ying, and to our children—Lara, Karl, Eva and Lillian—who patiently tolerated (less than or equal to) seven years of weekly stress.

Credits

The American Mathematical Society gratefully acknowledges the kindness of these institutions and individuals in granting the following permissions:

Bruce Bilney, `bruce@ozzigami.com.au`
Kangaroo tesselations; see pp. 3–5.

Creative Commons Attribution—Share Alike 3.0 Unported
Photo of Jessica Watson; see p. 43.

Reed Hutchinson/UCLA
Photo of Terrence Tao; see p. 47.

John Gollings Photography
Images of the Klein Bottle House; see p. 85.

McBride Charles Ryan
Physical model of the Klein Bottle House; see p. 88.

Monash University Archives
Newton apple tree; see p. 90.

Part 1

A day in Australia

If nothing is as American as apple pie, then nothing is quite as Australian as a mob of kangaroos. Except, perhaps, a 15-meter tall pineapple: for some reason, Aussies get a real kick out of biggifying things.

The following chapters dealt with scenarios that are typically, if not uniquely, Australian. The scenarios are merely gimmicks, playful and somewhat contrived excuses to discuss mathematics of independent interest. Nonetheless, they turned out to be pretty successful gimmicks, and the columns seemed to strike a chord.

CHAPTER 1

A puzzling Australia Day

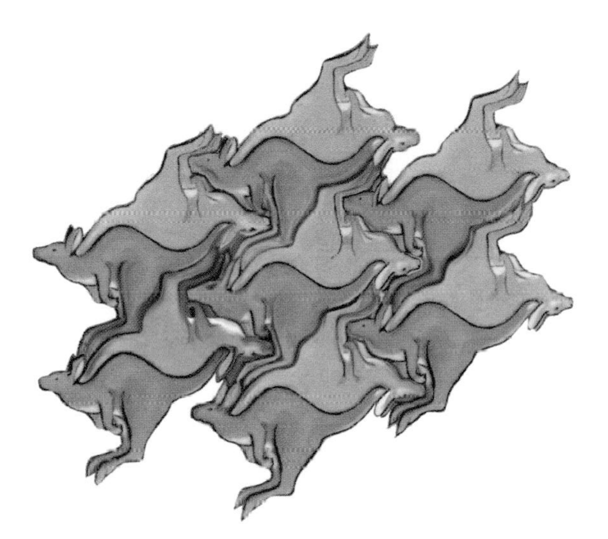

Last Australia Day a couple of patriotic Maths Masters met up to solve a 1000-piece jigsaw puzzle, a map of Australia. It was super-patriotic, since the map was covered with emus and echidnas and platypuses and the like.

We finished the puzzle, but only late in the night: our progress was slowed by the distracted pondering of jigsaw mathematics. We thought about the relationship between the number of pieces in a puzzle and the time to complete it. We wondered how the puzzle makers ensure that each piece fits in just one spot in the puzzle. But, the question that really slowed us down was this: *How simple can a 1000-piece jigsaw puzzle be?*

An easy example is the tiling of a bathroom, consisting of identical squares: even blindfolded, anybody could complete such a "puzzle". The same would be true if our tiles were equilateral triangles or regular hexagons.

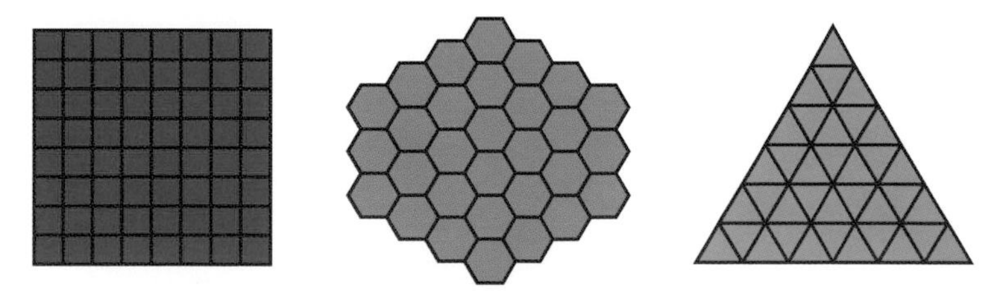

These three tilings with regular polygons are notable for their simplicity and symmetry, but no one is likely to be thrilled by a jigsaw puzzle consisting of 1000

red squares. There are other obvious choices of tile, such as rectangles and parallelograms, but they are hardly more interesting.

However, there are ingenious methods to transform simple tiles into complicated tiles in the form of real-life objects. These were pioneered by the brilliant graphic artist M. C. Escher.

Oddly, Escher the Dutchman never used cuddly Australian animals in his work. This is where South Australian artist Bruce Bilney comes in. The kangaroo tiling at the top of this chapter is his work, and below is another example: kangaroos interlaced with decorated Australias and tiny possum heads. For many more examples of Bilney's ingenious tilings, check out his Ozzigami website.

To see how Bilney creates his works, take a closer look at the tiling above. We have rotated it, and superimposed a square tiling. This demonstrates that Bilney's pattern can be regarded as a standard square tiling, but with each square decorated as shown on the right.

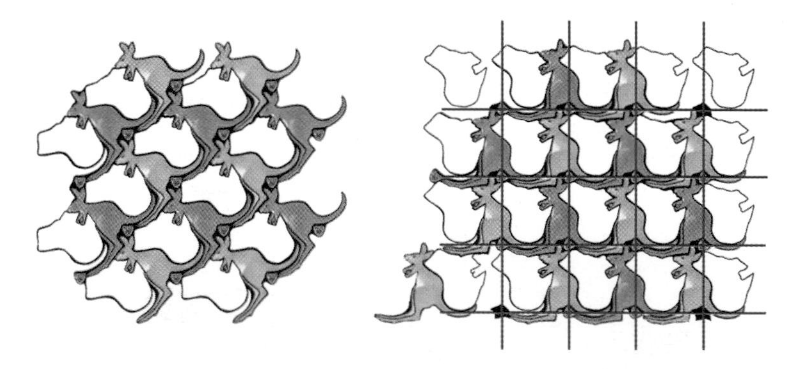

For the picture to tile properly, whatever sticks out from one side of the square has to match up with the other side. (Bilney has chosen to alternate red and grey kangaroos, which is why the colours don't match). But in fact that's all we have to worry about, and so we have an easy way to create interesting tilings: start with a square, cut out a couple of pieces from one edge, and reattach them on the opposite edge. The resulting shape can still be used as a tile.

By cutting out and reattaching more shapes, it is then easy to make complicated and striking tiles. Below, we have illustrated the steps involved in the making of our own animal tile: the fabled cone-headed camel from the outback. Not a bad creation, from just a few cuts and pastes.

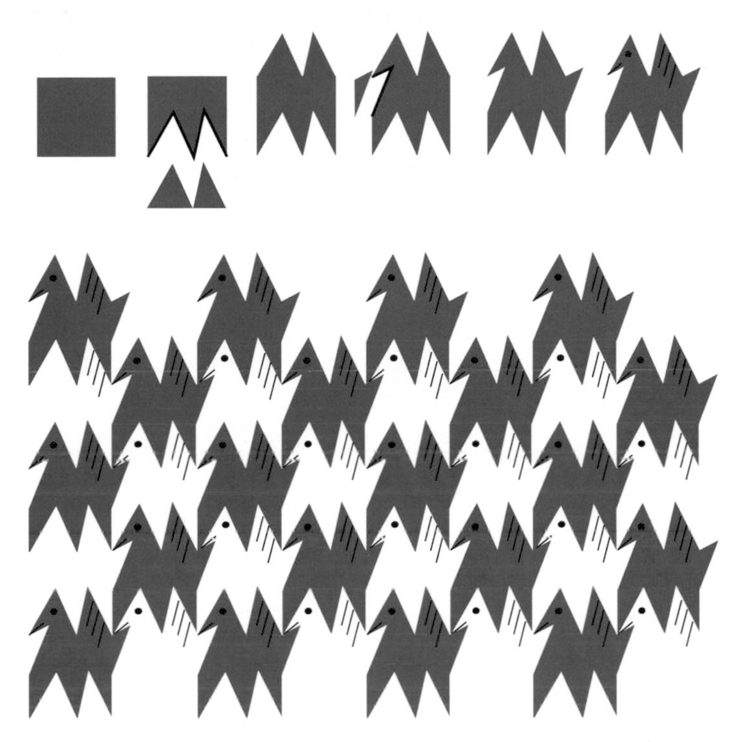

Puzzle to ponder

The rectangle superimposed on Bilney's masterpiece below is 16 centimeters wide and 13 centimeters high. What is the area of one of the kangaroos?

CHAPTER 2

The biggest pineapple in the world

There's nothing more Australian than to visit one of Australia's Big Things. This year, a Maths Master took his family on a pilgrimage to the Big Pineapple in Woombye, Queensland. It is big indeed, which pleased the kids. And, your Maths Master was also very pleased. We'll explain why.

The skin of the monster fruit is made of hexagons, stuck together as in a bathroom tiling and then wrapped to form a cylinder. Connecting hexagons in line, we can picture three different types of bands winding around the pineapple, as highlighted below. It turns out that the Big Pineapple has 13 green bands, 13 blue bands and 26 red bands.

You can also see such bands in real pineapples, but the numbers are different. It is well known that a real pineapple exhibits consecutive Fibonacci numbers, such as 5-8-13 or 8-13-21.

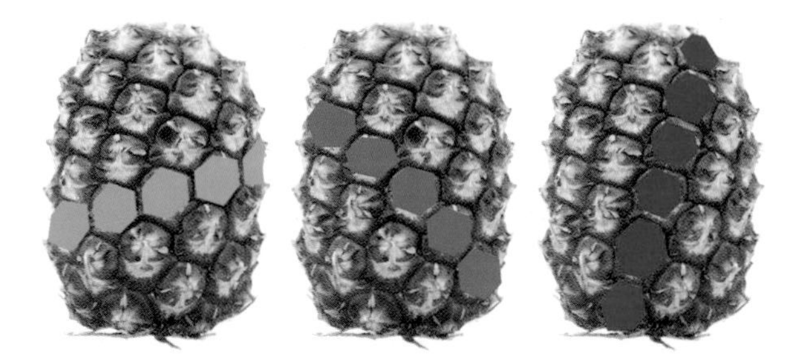

What is famously special about the Fibonacci numbers is that, starting with 1 and 1, each number is the sum of the two previous ones: $1 + 1 = 2$, $1 + 2 = 3$, $2 + 3 = 5$, and so on. Fibonacci numbers often occur in nature, and the numbers of spirals in pineapples and sunflowers are commonly cited examples.

All this is really fantastic. Except, a lot of it isn't true.

Recently, your Maths Masters have been spending a lot of time in greengrocers. We did indeed find a pineapple with Fibonacci numbers, the one with 5, 8, and 13 bands pictured above. However, we only found this pineapple after checking out 26 duds. Or, 27 if you include Woombye's Big Pineapple.

Is there something wrong with Australian pineapples? Should the food authorities order a massive recall? Probably not.

Fibonacci numbers are mathematically beautiful, but people tend to be carried away by their enthusiasm, and wishful thinking can sweep caution aside. So, a truthful claim that "some specimens of some plant species exhibit Fibonacci numbers" can morph into the very false declaration of a universal law. And, it is simply not the case that all or most pineapples exhibit Fibonacci numbers. Indeed, for many pineapples, the hexagons are so irregular as to make the counting of the bands practically impossible.

However, all is not lost. As a model for our pineapples, let's take a closer look at how a perfect cylinder can be tiled with perfectly regular hexagons. Cutting open and unwrapping the cylinder produces a rectangle, and then it is not too difficult to count the number of bands of each type. For example, the unwrapped cylinder below has 5, 8, and 13 bands: a good model for our Fibonacci pineapple above.

As the Big Pineapple demonstrates, the band numbers do not have to be Fibonacci numbers. Nonetheless, it is still possible to prove a very beautiful Fibonacci-like theorem:

Tile a cylinder with hexagons, and count the number of bands of each type. Then the largest number is the sum of the two smaller numbers.

For example, though Woombye's Big Pineapple is not Fibonacci, its band numbers do satisfy the equation $13 + 13 = 26$. In fact, given any three numbers with $A + B = C$, it is always possible to make a "pineapple" with those numbers of bands.[1]

So, expecting all pineapples to be exactly Fibonacci is too much to ask, but nicely grown pineapples with well-formed bands do still obey a simple and beautiful rule.

We'll finish by suggesting why the addition rule gives some hope of finding Fibonacci numbers in pineapples and other banded plants. (In fact, sunflowers and other plants tend to grow more regularly than pineapples, and they exhibit Fibonacci numbers much more predictably.)

Big pineapples grow from baby pineapples, and as they grow the numbers of bands change. What tends to happen at each stage (don't worry about the details!) is that the two largest band numbers will stay the same, and the new band number is the sum of these two.

This means that the band numbers in a baby pineapple tend to determine the future band numbers. And so, if our baby pineapple exhibits Fibonacci numbers, such as 2, 3, and 5, and if the pineapple grows nicely, it will continue to exhibit Fibonacci numbers.

But of course not all babies are so amenable. A quick calculation shows that Woombye's Big Pineapple must have been an extremely peculiar baby.

Puzzle to ponder

Why must Woombye's Big Pineapple have been a very strange baby?

[1] See Burkard's *Mathologer* YouTube video "The fabulous Fibonacci flower formula" for a detailed explanation: `youtu.be/_GkxCIW46to`

CHAPTER 3

The Nullarbor conundrum

The Nullarbor Plain is one of Australia's iconic geographical features, and taking the Indian Pacific railway across the Nullarbor is one of the world's great train journeys. The trip includes the world's longest straight section of railway line: 478 km without a bend.

The Nullarbor Plain is so flat and the railway line so straight, they are excellent models of the ideal planes and lines that mathematicians love. It was probably for this reason that the Nullarbor was chosen as the setting for a superb puzzle, which first appeared in 1975 in the collection *Chez Angelique* by J. Jaworski et al.:

> Edward Eyre Junior is lost on the Nullarbor Plain.[1] He hears a train whistle. Ed cannot see the train or the railway line, but he knows the direction from which the whistle came. His only chance of surviving is to run to a point on the line before the train gets to that point. In which direction should he run?

[1] Edward John Eyre was a famous 19th-century explorer of Australia. In 1841, Eyre and Wylie, an aboriginal guide, travelled 1000 miles across the Nullarbor Plain to the west coast of Australia.

We can treat this as a purely mathematical puzzle, but let's first consider a real-live Ed on the real Nullarbor. Then the railway line is basically running east-west. We'll also assume Ed can read the Sun or the stars well enough to know the direction of north. Then the scenario would be as pictured below:

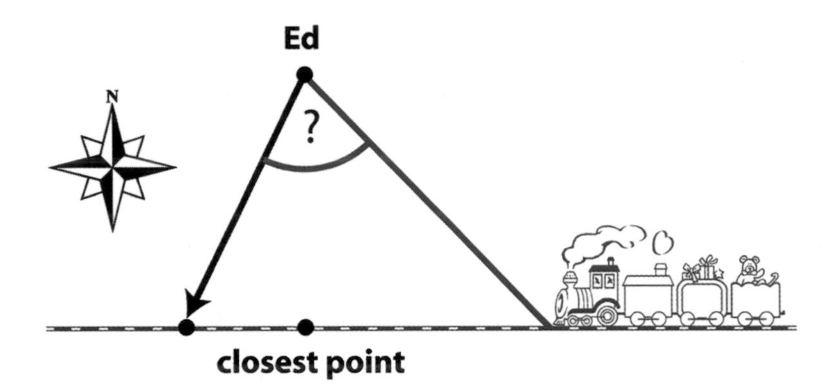

The closest point on the rail line is due south of Ed. Note that Ed may as well assume that the train is travelling towards this point. Otherwise, unless Ed can outrun the train, he is doomed no matter in which direction he runs.

But this still leaves the question of the direction in which Ed should run. One would imagine that the correct choice depends upon the positions of Ed and the train, and of their speeds. Surprisingly, this is not the case.

Applying some simple trigonometry or similar trickery, it turns out that all the factors of distance and speed cancel: the very elegant solution is that Ed should always run at right angles to the original direction of the whistle. He may or may not be able to run fast enough, but if he can in fact save himself, that is the direction to go. We'll leave the details for you to puzzle over.

But what now if the puzzle is purely mathematical, with an Ideal Ed lost on an ideal Nullarbor PLANE? We'll still assume the train is travelling along a straight line, but we no longer have compass directions to guide us. Does this change the puzzle? What's the best advice for Ideal Ed to attempt to save himself?

Reference: *Chez Angelique: The Bumper Late-night Problem Book*, Chez Angelique Publications, 1975, 64 pages.

Puzzle to ponder

Can you show that Real Ed should travel at right angles to the direction of the train whistle? How does this change for Ideal Ed?

CHAPTER 4

The revenge of the lawn

The grass is long, and guess whose job it is to cut it? Confronted with this tedious chore, a Maths Master has only one possible course of action: he must immediately retire to his office to determine the mathematically best way to mow the lawn.

Clearly, what is optimal for a lazy Maths Master is to push the lawnmower the shortest distance possible. So, he wants to avoid mowing twice over the same patch of lawn. Let's see what he might have to consider.

As we push the mower, its circular slashing disc cuts a swath through the grass. If we travel in a straight path, then the cleared region will be a corresponding thick line. Think of the straight path as consisting of four shorter segments. Now, rotate somehow at the three joins to create a new, crooked path.

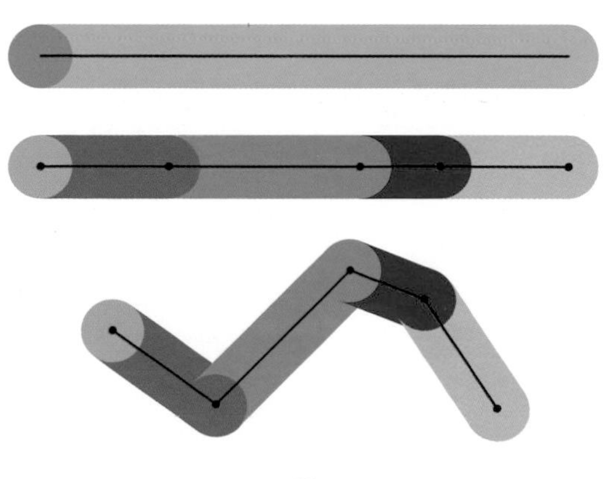

Of course the straight and crooked paths are of exactly the same length. However, the crookedness creates some extra overlap at each junction, and so the area of the surrounding mown region will be lower. And, the sharper the angles, the greater this wasted overlap.

It is therefore beneficial to try to mow along paths that are as straight as possible. However, before someone declares the problem solved, and especially before they hustle us out to do some mowing, we need to think about how to determine such "straight paths" for a whole lawn.

It is also worth mentioning that such analysis has other applications almost as important as improving mowing technique. For example, consider a helicopter searching for a person lost at sea. The helicopter should survey as much ocean as quickly as possible. And, at any moment, the search team can survey a circular patch of ocean surrounding the helicopter. So, this search problem is mathematically similar to our problem of mowing as efficiently as possible.

Of course, the practical concerns of the two problems are very different. When searching for someone lost at sea, time is of the essence, and any overlap is a serious issue. As well, searching over a large, uninterrupted body of water, we can quickly decide upon a suitable search: for example, begin where the person was last sighted, and then search in an expanding spiral path.

On the other hand, suppose we have our lawn in front of us and we want to determine a good path. In practise most lawns are small and quite regular in shape. So, a little trial and error usually suffices to come up with a suitably efficient mowing path. However, if we need to mow a golf course, the problem is trickier.

For most large (and more realistic) optimizing problems, the trick is to make them manageable by simplifying them and by being satisfied with a close-to-best solution. Then, practical solutions will come from a computer.

For example, we can simplify our golf course mowing problem by covering the course with an array of circles, as pictured, with each circle radius equal to the width of the mower disc.

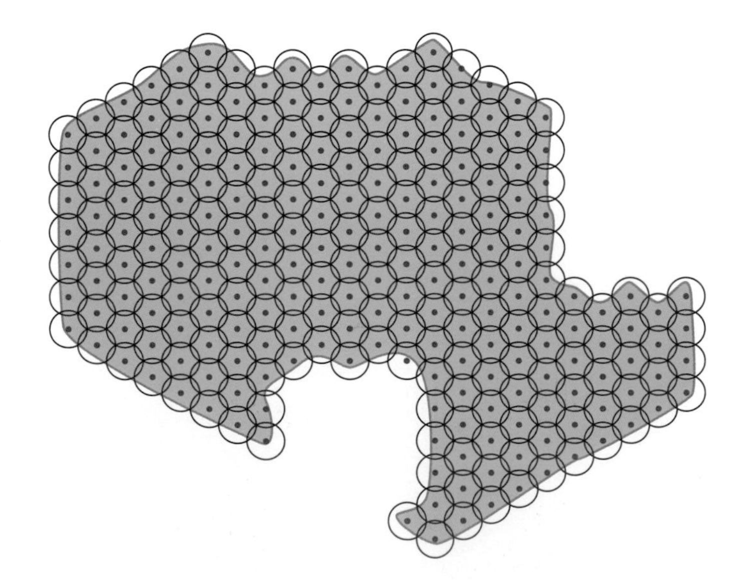

Connecting the centres of the circles produces an equilateral triangular grid, with vertices at the circle centres. Following a path consisting of grid edges, there will necessarily be a fair amount of overlap.

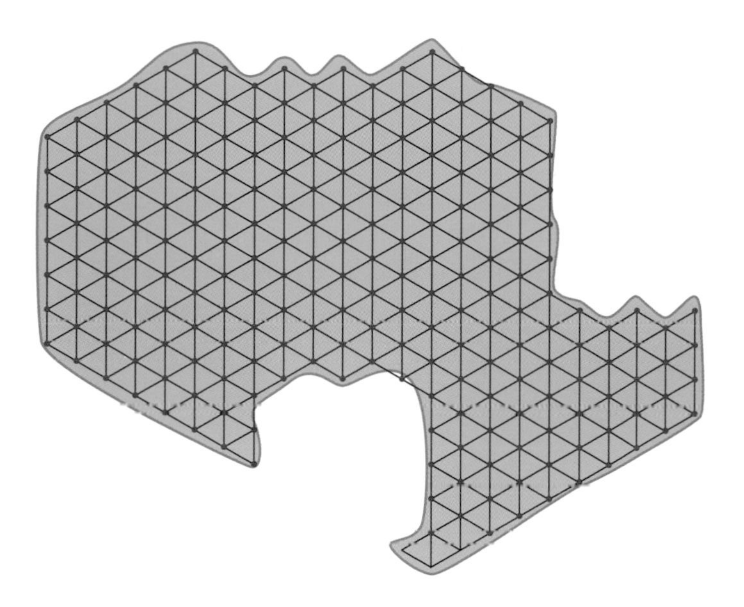

The advantage is, as long as all vertices are included in the path, we can be sure that the whole lawn will be mowed. As well, the efficiency of a path no longer depends upon straightness and is measured simply by the number of edges travelled: the best we can hope for is to travel along one fewer edge than the number of vertices.

So, a path will be relatively efficient if few or no vertices are visited more than once. Finding such an efficient path is easily achieved by well-known computer search algorithms.

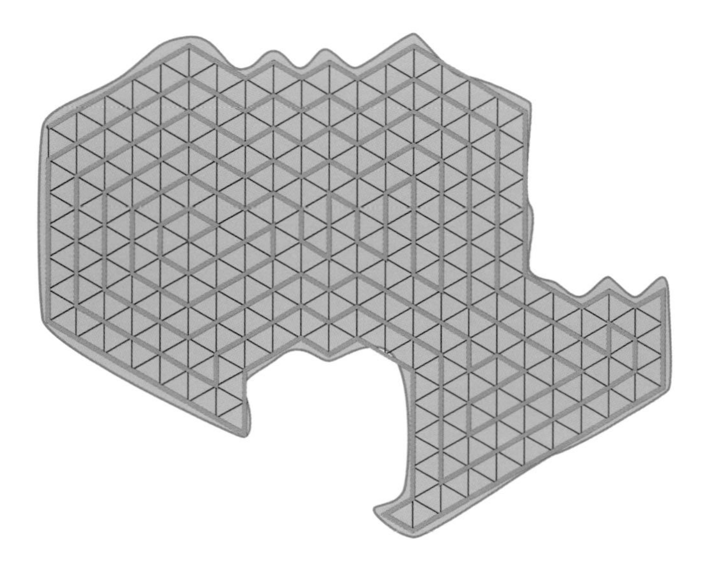

So, we now have an elegantly mathematical approach to mowing the lawn. But, alas, that analysis took a while, and it seems to be getting dark. The mowing will just have to wait until next weekend.

Reference: Versions of the following interesting technical article are freely available on the internet: E. Arkin, S. Fekete, and J. Mitchell, *Approximation algorithms for lawn mowing and milling*, Comput. Geom. 17 (2000), no. 1-2, 25–50.

Puzzle to ponder

Suppose you have a lawn that is 200 square meters in area and your mower disc is 80 centimeters in diameter. What is the minimum length of the path you'll have to mow?

CHAPTER 5

A Greek in an Italian restaurant

Melbourne is a wonderful Greek city. Greek? Yes, Greek. By pretty much any measure, Melbourne has the largest Greek population of any city in the world outside of Greece. So, it shouldn't be all that surprising to encounter the occasional homage to Pythagoras, one of the ancient world's mathematical superheroes. However, we were very surprised to find Pythagoras appearing in our local Italian restaurant.

Entering the restaurant, we looked straight past the menu to the unusual tiling on the floor. It consists of small grey squares and larger brown squares. This may not seem so Pythagorean, but the triangles really are there, just slightly hidden.

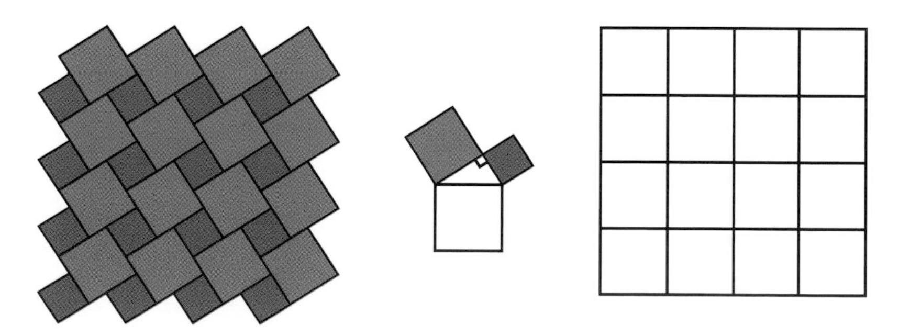

In the diagram above we've employed the sides of the grey and brown squares as the smaller sides of a right-angled triangle; we've also drawn an empty square off of the hypotenuse. Pythagoras's theorem then says that the sum of the areas of the grey and brown squares is equal to the area of the empty square.

What is remarkable is that hidden in the floor tiling is a *proof* of Pythagoras's theorem. The left diagram above is a patch of the floor tiling consisting of 16 squares of each colour. The right diagram is composed of 16 of the large empty squares in a grid. The two diagrams can now be naturally superimposed, but with some bits of the coloured squares sticking out.

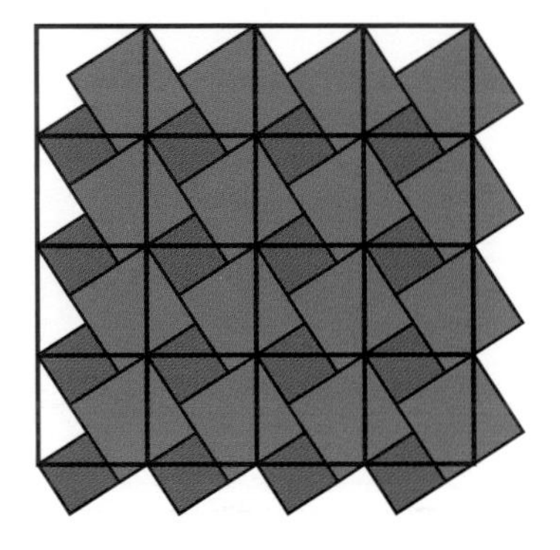

Here is the punch line: all the coloured bits sticking out on top and to the right can be cut off and exactly refitted into the spaces down below and on the left. This means that the 16 empty squares and the 32 coloured squares sum to exactly the same area. Divide by 16, and we have Pythagoras's theorem! Q.E.D.

But have we really proved Pythagoras's theorem? What about other right-angled triangles? Or what if we had used a tiling with squares of different sizes? Well, the final question answers the first two. Starting with any right-angled triangle, we can use the two short sides to make a restaurant tiling with coloured squares. Then we create the grid of empty squares, and they fit as hypotenuse squares just as before. We argue just as above, and Pythagoras's theorem holds true again.

By the way, the brown squares in our restaurant are exactly twice the size of the grey squares. This means that the hidden triangles have exactly the same proportions as the triangles in the façade of Melbourne's Federation Square.[1] Definitely the most mathematical restaurant in Melbourne.

Puzzle to ponder

What special case of Pythagoras's theorem is hiding in an ordinary bathroom tiling consisting of identical squares of alternating colour?

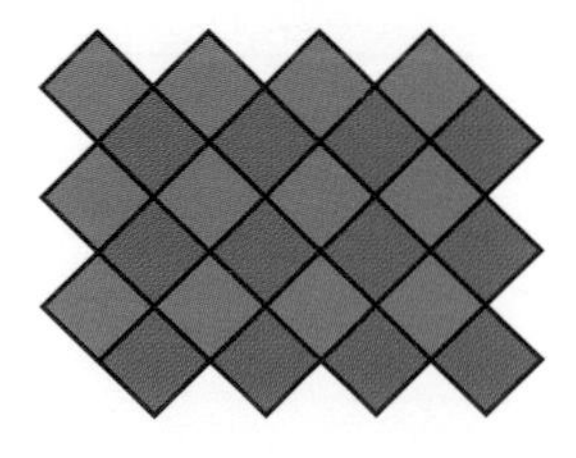

[1]See Chapters 18 and 19.

CHAPTER 6

Please say hello to Adelaide and Victoria

We would like to introduce you to two very pretty and very clever ladies. Above, we have Adelaide on the left and Victoria on the right. They are Venn diagrams.

Now if you recall something of Venn diagrams from school, you may be puzzled: aren't there only two such fellows, simple combinations of circles, and bearing no resemblance to our ladies above? Not true! There are in fact infinitely many Venn diagrams.

So, what is a Venn diagram?

Let's start with a large collection, say the collection of all people in Australia. Mathematicians use the technical word *set* to refer to such a collection. It doesn't really matter here, except that "collection" suggests that we've physically gathered all the people together in our backyard. Talking of sets emphasises that we're only doing the grouping in our minds.

We now consider two *subsets* (that is, smaller collections) of the set of Australians: the subset of Australians who like to read, and the subset of Australians who like math. So, an Australian is either a reader or not, and they're either a math fan or not. That results in Australians being classified into a total of $2 \times 2 = 4$ types.

A Venn diagram is simply a pictorial representation of such a classification. The whole box represents all Australians, and the two circles represent the readers and the math fans. The four smaller regions created by the overlapping circles then correspond to the four types of Australians in our classification.

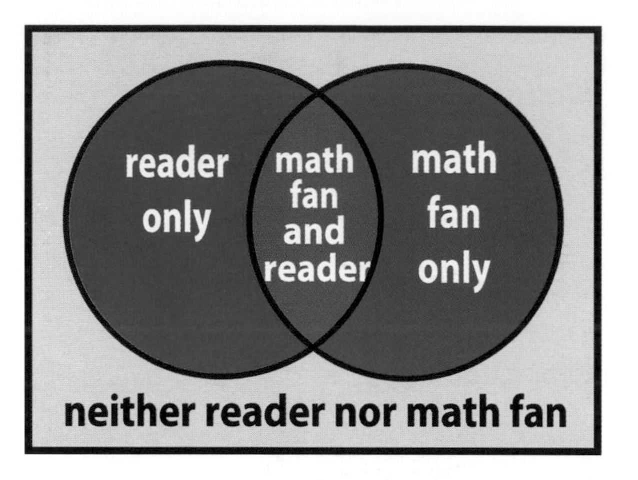

Suppose we now consider a third subset, the set of Australians who love cats. There are then $2 \times 2 \times 2 = 8$ different ways in which a person can belong or not belong to our three subsets. Representing the cat lovers by a yellow circle, the types of people again correspond to the (eight) regions created by the overlapping circles.

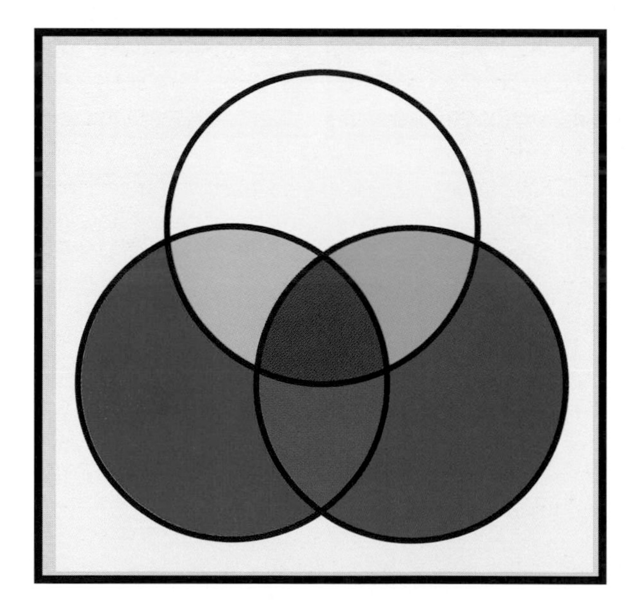

Why stop here? Let's consider the subset of Australians who hate calculators, and also the subset of Australians who are unimpressed by the National Curriculum. We now have $2 \times 2 \times 2 \times 2 \times 2 = 32$ different types of Australians. But, how do we represent this pictorially?

The following elegant diagram consists of five ellipses, whose overlaps create a Venn diagram consisting of 32 regions. This enables us to represent our 32 types of Australians. You can probably guess where your Maths Masters are located.

Can we go further? The 19th-century mathematician John Venn was the first person to systematically investigate such diagrams. He was able to show that a suitable diagram can be drawn based on any number of subsets.

There are in fact many different methods of constructing Venn diagrams, but the diagrams pictured above are the prettiest and most symmetric. The charming Adelaide and Victoria are Venn diagrams based on seven subsets, each composed of 128 regions. They are each constructed by taking a single loopy curve, and overlapping 7 rotations of that curve. The result is a diagram with *7-fold rotational symmetry*. Similarly, the ellipse diagram has 5-fold symmetry.

It turns out that the rotationally symmetric Venn diagrams can only be drawn if we begin with a prime number of subsets. And, these diagrams are not easy to find. The symmetric diagram for 11 subsets was only discovered in 2002. Then, in 2004, mathematicians Jerrold Griggs, Charles Killian, and Carla Savage discovered a general procedure that works for any prime number of subsets.

And what about the names? Well, Adelaide was (re)discovered and named by mathematician Anthony Edwards, while he was visiting the University of Adelaide in Australia. And, Victoria was discovered by Frank Ruskey at, of course, the University of Victoria (in Vancouver, Canada).[1]

Puzzle to ponder

1. It turns out to be impossible to draw a Venn diagram for four subsets using four circles. What about using curves other than circles?

[1] A geographical pun: "Victoria" is also the name of a state in Australia.

2. In each region of the Venn diagram below, we have indicated how many ellipses cover that region. You can check that there are, respectively, 1, 5, 10, 10, 5, and 1 regions in which 0, 1, 2, 3, 4, and 5 ellipses overlap. Where do these numbers come from? What would the corresponding numbers be for Victoria and Adelaide?

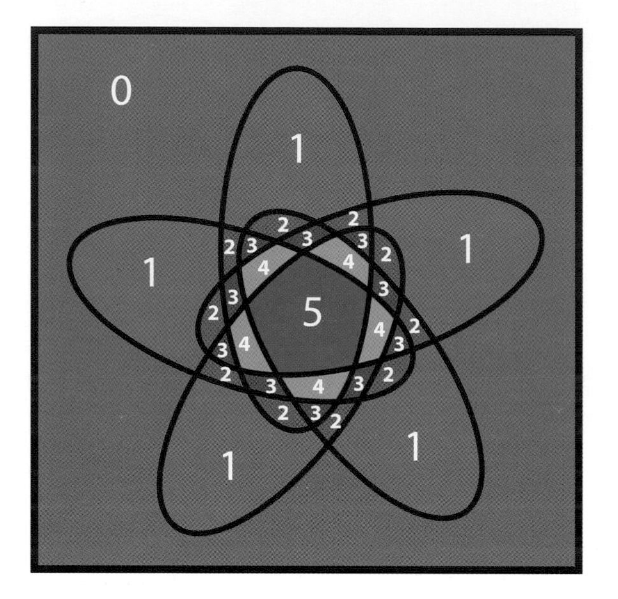

Part 2

Sports rules!

Australians love sports. They may take little notice of high achievers in mathematics and science, but Aussie sports stars are almost guaranteed to become national heroes. And of course all that sports brings with it a trove of statistics for us to ponder. Just as baseball exists in a cloud of statistics, so does cricket and rugby and "Aussie rules" football. It all provides plenty of material for mathematics articles with a popular hook, and with almost no deeper meaning.

Most sports that Australians watch and play will be quite obscure to an American audience. However, in the main the finer details don't matter all that much: a ball is hit/kicked/thrown, resulting in points being scored, and that's about it.

CHAPTER 7

The problem of cricket points

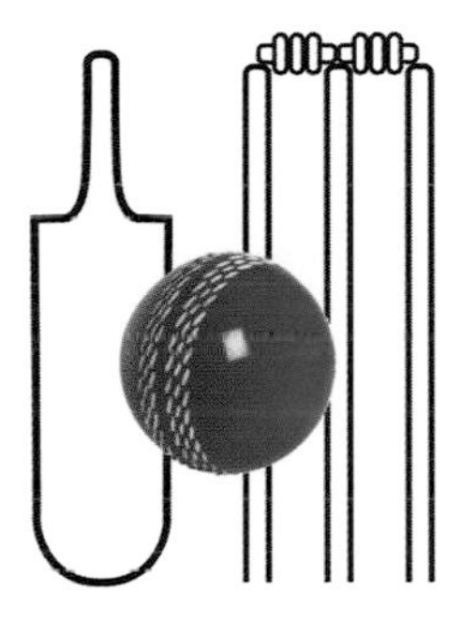

If a game is interrupted, how do we decide the winner? We could simply declare the game a draw, but if one team is well in front, they could justifiably feel this is unfair. And in professional games there are often strong administrative or financial incentives for declaring a winner.

In our modern world the notorious scenario for this question is one-day cricket, and what to do when rain interferes.[1] Equally notorious is Duckworth-Lewis, the currently sanctioned method for adjusting the game. We'll try to make D-L less notorious by giving a simple explanation of how it works.

Let's imagine Australia (yay!) and India (boo!) are playing a match, Australia bats first and scores 300 runs in their 50 overs.[2] Then the rains come down, and the match is shortened to 30 overs each. How many runs should India have to score to win?

The simplest approach is to take Australia's average rate of 6 runs per over, and then multiply that by 30. So, India would have to score 180 runs to tie, with one more to win. But this would obviously be unfair to Australia. With only 30 overs, India's batsmen could afford to be more reckless, making their task easier. So, India should have to score more than 180 runs, but how many more?

---

[1]Cricket in Australia (and elsewhere) plays a similar cultural role to baseball in America: it's a slow game with a proud and dusty history, with lots of numbers and relatively little hoopla. The traditional form of "test" cricket is played over five days, with a game often ending in a draw. (And you thought baseball was slow.) Since the 1970s, a one-day version of the game has become very popular, and it's this form of the game we're writing about. Since this article was originally published, a shorter "T20" version of the game has grown in popularity; the issues we discuss in this chapter also apply to this shorter version.

[2]A one-day game consists of one inning for each side, each inning consisting of 50 *overs* and each over consisting of 6 *balls* (pitches). Each team consists of 11 players, all of whom bat. Once 10 players are out (called a "wicket"), the team's inning is over, whether or not the 50 overs have been completed.

There is no definitive answer. We just have to look at the statistical evidence of how teams score their runs. Then, any method of adjustment will be a compromise between accuracy and simplicity. Duckworth-Lewis, first used in 1997, is such a compromise.

The D-L approach is to consider the "resources" of each team, consisting of their 10 wickets and their 50 overs. In the example above Australia used all their resources, but India had some of theirs taken away. To estimate how much, Duckworth and Lewis modeled how teams score (they used exponential functions) and created a table like the one below.

	Wickets lost			
Overs left	0	2	5	7
50	100	85.1	49	22
40	89.3	77.8	47.6	22
30	75.1	67.3	44.7	21.8
20	56.6	52.4	38.6	21.2
10	32.1	30.8	26.1	17.9

Initially, a team has 100% of their resources, but this declines as the overs are used and the wickets fall. For example, with 40 overs left and 7 wickets lost, the table indicates that only 22% of the team's resources are left.

With this approach, we have an easy method of analyzing games like the one above. India has 10 wickets and 30 overs, which amounts to 75.1%, just over 3/4, of their resources left. Since Australia had 100% of their resources, India should only have to make 3/4 of Australia's runs. Then, 3/4 of 300 is 225 and so that is their target, with 226 to win.

D-L is not perfect, but it is relatively simple and very flexible. For example, suppose in the next game Australia again bats first and again scores 300 runs. Then India scores 150 runs in their first 30 overs, without losing any wickets, and then the rains come down. Who should be declared the winner? We'll leave this for you to puzzle over.

But is D-L fair? The evidence is that it works very well. In 2004 Duckworth and Lewis looked at the effect of their system, analyzing over 1000 games. They calculated that in completed games, the team batting first won about 52% of the time. And, in games where D-L was applied, the team batting first won about 53% of the time. This is very good agreement!

Still, Duckworth and Lewis weren't satisfied. They had evidence that their system did not work as well when one team made a huge score, so they adjusted their method. Now, tables are no longer sufficient, and the calculation is done by computers. And D-L now works even better. But, is it cricket?[3]

Puzzle to ponder

Who won our game above, where Australia scored 300 runs, and then India scored 150 runs in 30 overs with the loss of no wickets? What if India had lost two wickets?

[3]An English expression, questioning whether something is fair. Cricket was traditionally, and is no longer, played with a very high degree of sportsmanship.

CHAPTER 8

Matthew Lloyd vs. Brendan Fevola

Mathematics is not merely beautiful and fun, it is also a powerful tool to answer very important questions. Such as: who is the better full forward, Matthew Lloyd or Brendan Fevola?[1] Answering such a question is not only important for bragging rights, it can also influence critical decisions on who to include in a fantasy football team.

So, who is the better full forward? In 2007 Lloyd and Fevola both played 19 games, with Lloyd's 62 goals edging out Fevola's 59. In 2006, Fevola played many more games than the injured Lloyd, and it makes more sense to compare their average goals per game played. By this comparison, Lloyd again had the edge, averaging 4.3 goals per game to Fevola's 4. It seems clear, at least from goal averages in the last two years, Lloyd has been the superior spearhead.

The goalkicking averages for Matthew Loyd and Brendan Fevola

	2007	2006	2005
Lloyd	62/19	13/3	59/20
Fevola	59/19	84/21	49/19

But now we can make a very puzzling observation. Suppose we consider 2007 and 2006 combined. Over the two years, Lloyd kicked 75 goals in 22 matches, for an average of 3.4 goals per game. And, by comparison, Fevola kicked 143 goals

[1] In Australian rules football goals can be scored by any player, however the full forward is the player with the greatest responsibility for kicking goals.

over 40 games, for an average of 3.6 goals per game. So, though Lloyd had a better average in each year, Fevola had a better average overall!

In fact, if we include the data from 2005, the puzzle continues. In 2005, Lloyd again averaged more goals per game than Fevola. But, over the years 2005 to 2007 combined, Fevola still had the higher average.

This strange phenomenon—comparing averages of separate portions of some data contradicts an overall comparison—is known as *Simpson's paradox*. The possibility of this occurring was first recognised over 100 years ago, and there are now striking examples known in many real-world areas: sport of every kind, tuberculosis infection rates, sex discrimination claims, and on and on.

In fact, in a sense, wherever there is data, there is Simpson's paradox. In a recent paper, the mathematicians O. E. Percus and J. K. Percus proved that essentially any set of data can be cunningly split in a way to give rise to Simpson's paradox.

At first glance it seems very peculiar, but Simpson's paradox is actually easy to explain. In our goalkicking example, the key is the year 2006. In that year both Lloyd and Fevola had a high average, with Lloyd's slightly higher. However, Lloyd only played 4 games, which had little effect on his long-term average. On the other hand, Fevola's sustained form over 21 games had a significant effect.

Undoubtedly, averages can be meaningful and informative. So, what's the moral of this story? Don't let a high average over a short period mislead you. And, there's only one Tony Lockett.[2]

Puzzle to ponder

What is the lowest number of goals Fevola could have kicked in his 19 games in 2007 and still have his combined average over 2006 and 2007 be greater than Lloyd's?

[2]Don't ask. OK, since you asked, Tony Lockett was perhaps the all-time greatest full forward. A song is often sung about him, *There's only one Tony Lockett*, to the tune of *Guantanamera*.

CHAPTER 9

The perfect rugby conversion

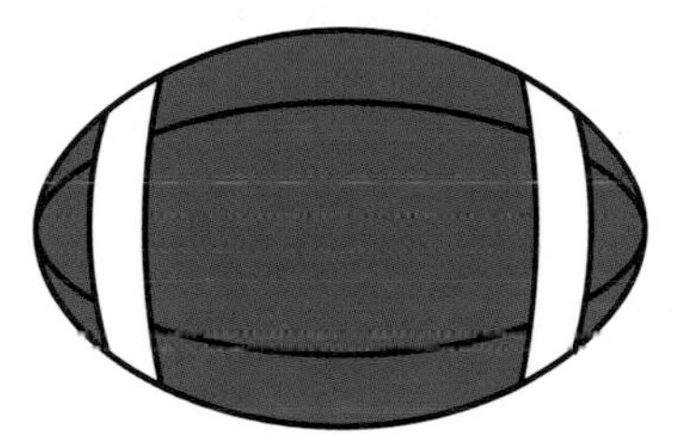

Yes, we know. Melbourne is an Aussie Rules town. But now we can all become passionate rugby league fans.[1] Australia is set to avenge their shock loss to arch rival New Zealand in last year's World Cup.

We've been discussing tactics with Cameron Smith, star kicker for the Melbourne Storm. To give the Kangaroos an extra edge against the Kiwis, we are figuring out a Maths-Masterly method of kicking goals.

As a reminder, a player scores a try by grounding the ball at a point in the *in-goal area* (that is, the end zone), beyond the opposition's goal line. The kicker then places the ball wherever he wishes along the red line pictured, through the grounding point and perpendicular to the goal line: we'll call this the conversion line. To score the extra points for the conversion, the ball must then be kicked between the goal posts and above the crossbar.

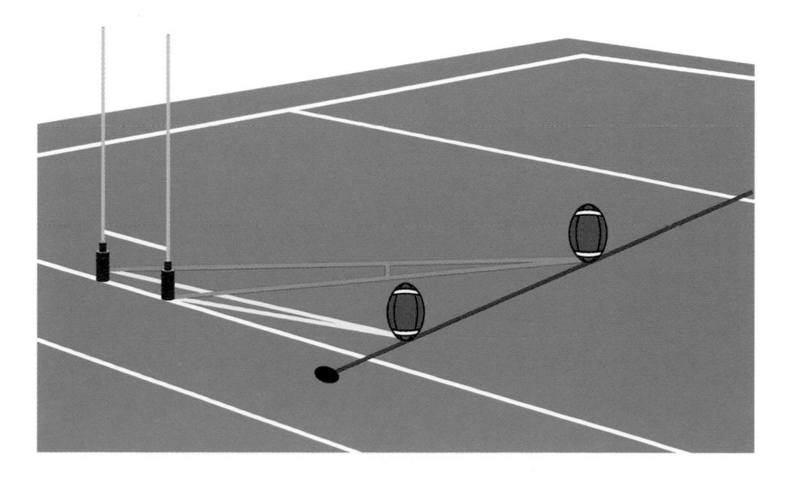

[1] This column was written in 2009. Rugby is a precursor to gridiron and is still popular in many countries, including parts of (but not all of) Australia. A *try* corresponds to a touchdown, and a *conversion* corresponds to an extra point.

But where should the kicker place the ball? If the ball is grounded between the posts, then the decision is pretty much a no-brainer. The kicker simply comes close to the goal line, subject to allowing enough distance for the ball to rise above the crossbar.

The method we describe deals with the trickier situation, when the try is scored to one side of the goals. In this scenario, the kicker should allow as much leeway as possible for skewing the kick. This amounts to making the angle subtended by the goalposts as large as possible.

But where is this spot? The angle is obviously zero at the goal line and then increases as we walk along the conversion line. Once we get to the other end of the field, the angle is again very small. The spot we're after is along the way, right where the angle begins to decrease.

A clever way to find the optimal spot is to draw circles passing through the bases of the two goal posts. One of these circles just touches the conversion line. That touching point turns out to be the optimal spot from which to kick.

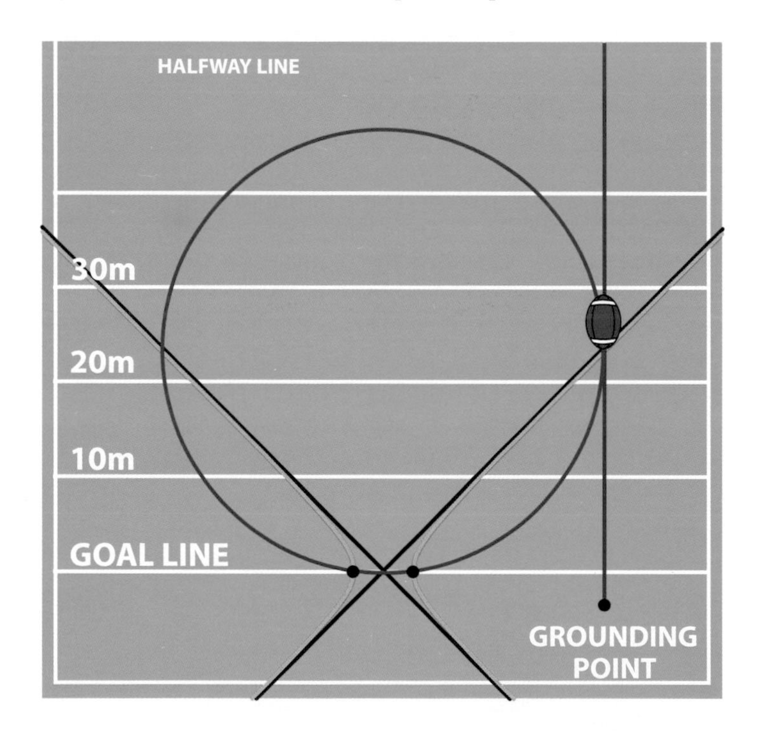

OK, it may be too much to expect the Kangaroos to come out drawing circles with their giant compasses. But there's another way.

If we take the optimal kicking spot from each conversion line, they combine to form the orange curve pictured. This curve should look familiar, since it is exactly a hyperbola. It is the graph $y = 1/x$, just rotated and scaled to fit on the field.

So, now we have rugby league players wandering out with their graphics calculators? Not necessarily. Notice that the hyperbola sits very close to its asymptotes, the crossing black lines.

These asymptotes hit the sidelines at about the 34 meter mark. So, the kicker can simply start there on the sideline and march straight towards the goals until he gets to the conversion line. That is then very close to his optimal spot. Simple!

There is one wrinkle to our calculations above. We have been drawing all our graphs down on the field, but the ball will hopefully be flying high over the crossbar. This changes the precise angles we need to investigate. Somewhat surprisingly, the optimal kicking spots are still along this same hyperbola.

We're finally done. And now what does Cameron Smith think of our brilliant mathematical plan? Alas, he simply prefers to kick from closer in, naively guided by factors such as deviation by the wind and curve of the ball in flight. And his intuition comes from having kicked a mere 341 career goals.[2]

We suspect Cameron knows exactly what he's doing and can comfortably beat the Kiwis without our help. It's back to the drawing board for the Maths Masters.

References: The following article, available at `www.qedcat.com/articles`, has all the gory details together with a comprehensive list of references:

B. Polster and M. Ross. *Mathematical Rugby*, Mathematical Gazette 94 (November 2010), issue 531, 450–464.

The article in which this mathematical way of locating the optimal conversion spot was highlighted for the first time is

Hughes, *A.*, *Conversion attempts in rugby football*, Mathematical Gazette 62 (1978), 292–293.

Puzzle to ponder

The method described above is based upon the following fact. Draw a circle passing through the two goal posts, pick a point on the circle, and draw lines to the bases of the two goalposts. Then, no matter the point, the angle formed by the two lines will be the same. Can you now fill in the details of the described method?

[2]Cameron Smith had kicked his 341 goals at the time of writing, in 2009. By mid-2017, this had risen to 1049 goals.

CHAPTER 10

The Socceroos and the group of death

It's soccer time. The World Cup begins on Friday, and we're all hoping for great success for the Socceroos.[1] They have a tough first round match against Chile, but then... Oh, no! Australia is in the group of death!

Well, actually, any group would be the group of death for Australia. They are currently ranked 62nd in the world, just below soccer powerhouse Burkina Faso, and are the lowest ranked team in the tournament. Unless someone quickly locates Harry Kewell[2] and dunks him in a fountain of youth, the Socceroos are in for a very torrid time.

It's not clear who is to blame for the decline in Australian soccer. (Your Maths Masters suspect Tony Abbott.[3]) However we've been puzzling over a different question: how exactly do we determine which is the group of death?

As a reminder, the World Cup will begin with eight groups of four teams; Groups A and B are pictured above, and Groups C to H are below. Each group plays a round robin, with the top two teams qualifying for the knockout phase of the tournament. So, the first and most important task for any team is to survive the group stage.

We've noted the world ranking of each team (as of June 5, 2014), giving some indication of the difficulty of each group. There's no question that having to face Spain, Chile, and the Netherlands is a tough ask, and it's not surprising that Australian sports journalists were quick to declare Group B the group of death. However, some less Aussie-inclined observers have opted for Group D, and Group G has also been labeled the group of deadliest deathly death. There has even been some deadliness detected in Group A. All in all, that's quite an abundance of death.

[1] This column was written just before the 2014 World Cup.

[2] Australia's star winger for the previous decade.

[3] Australia's prime minister at the time. Conservative, cruel and inept, he was aptly summed up by John Oliver as a "car crash of a human being". We made fun of Mr. Abbott (say it quickly) at every opportunity, and if there was no opportunity, then we made one up.

C	
COLOMBIA	8
GREECE	12
IVORY COAST	23
JAPAN	46

D	
URUGUAY	7
ITALY	9
ENGLAND	10
COSTA RICA	28

E	
SWITZERLAND	6
FRANCE	17
ECUADOR	26
HONDURAS	33

F	
ARGENTINA	5
BOSNIA	21
IRAN	43
NIGERIA	44

G	
GERMANY	2
PORTUGAL	4
USA	13
GHANA	37

H	
BELGIUM	11
RUSSIA	19
ALGERIA	22
SOUTH KOREA	57

Can we employ some mathematical method to determine the true group of death? No, of course not. However it's a lot of fun pretending that we can.

One simple idea is to just sum the rankings of the teams in each group; a low sum indicates the presence of strong teams, and so we might just pronounce the death of the group with the lowest sum. This method suggests that Group D is the toughest, with a total of 54, followed closely by Group G. With this ranking, Group B is only the fifth most deadly (or third liveliest), coming after E and C. That can't be right; surely we can "prove" Australia's group is deadlier than that.

There is something intrinsically dodgy about adding rankings as we have; it's definitely not the same as counting physical objects or measurements, such as carrots or kilograms. So, we might choose to work with some more tangible measure of teams' strengths—the number of matches won or the like. However, that is much easier said than done, with international teams having dramatically different schedules, playing opponents of dramatically differing abilities.

FIFA, the governing body of world soccer, attempts just such a calculation. By means of an astonishingly complicated formula, every match result is assigned a value, and the values are then weighted and summed to determine the total points for each team. The points table then determines the world rankings. These points are still not carrots or kilos but it's probably as close as we're going to get.

A	
BRAZIL	1242
CROATIA	903
MEXICO	882
CAMEROON	558

B	
SPAIN	1485
CHILE	1026
NETHERLANDS	981
AUSTRALIA	526

C	
COLOMBIA	1137
GREECE	1064
IVORY COAST	809
JAPAN	626

D	
URUGUAY	1147
ITALY	1104
ENGLAND	1090
COSTA RICA	762

E	
SWITZERLAND	1149
FRANCE	913
ECUADOR	791
HONDURAS	731

F	
ARGENTINA	1175
BOSNIA	873
IRAN	641
NIGERIA	640

G	
GERMANY	1300
PORTUGAL	1189
USA	1035
GHANA	704

H	
BELGIUM	1074
RUSSIA	893
ALGERIA	858
SOUTH KOREA	547

Summing the points in each group suggests that Group G is the deadliest, with Group D not far behind. Australia's Group B now comes out third toughest, so we're getting there. We'll see if we can do a little better.

When trying to determine the group of death, the question has to be asked: death for whom? Group B is somewhat less deadly exactly because it includes Australia, but that doesn't alter the fact that it's a nasty group for Australia. Indeed, the points of Australia's three opponents sum to 3492, second highest in the tournament; Ghana, with their opponents' points totalling to 3524, has the worst of it. That largely substantiates the Socceroos' death claims and is probably as far as we can go in that direction.

Now Australia, you probably think this chapter's about you (as Carly Simon might have sung). However, one does not determine groups of death according to whether cannon fodder like Australia have a tough draw. The real question is whether strong teams, those we would typically give a good chance in the group phase, have a notably difficult draw.

With this criterion, it is easy to do the sums to demonstrate that Groups D and G have the strongest death claims. In each group the top three teams are very strong, creating an intense competition for the two qualifying spots; any of those teams might feel death hovering near. In Group B, Chile and the Netherlands can also be a little grumpy; however, Spain shouldn't have a particularly tough time getting through.[4]

That was all a lot of fun but it was just simple arithmetic. Can one make the analysis more sophisticated? Absolutely, and many people love to do so. However, the fancy mathematics pretty much arrives at the same conclusions—some great teams will be playing some great do-or-die matches, and, whatever else, Australia is doomed.

Puzzle to ponder

Using FIFA's points for each country as a guide, what would have been the fairest way to sort the countries into eight groups?

[4]And of course Spain failed to do so. Australia was also eliminated at the group stage, but was more competitive than expected.

CHAPTER 11

The greatest team of all

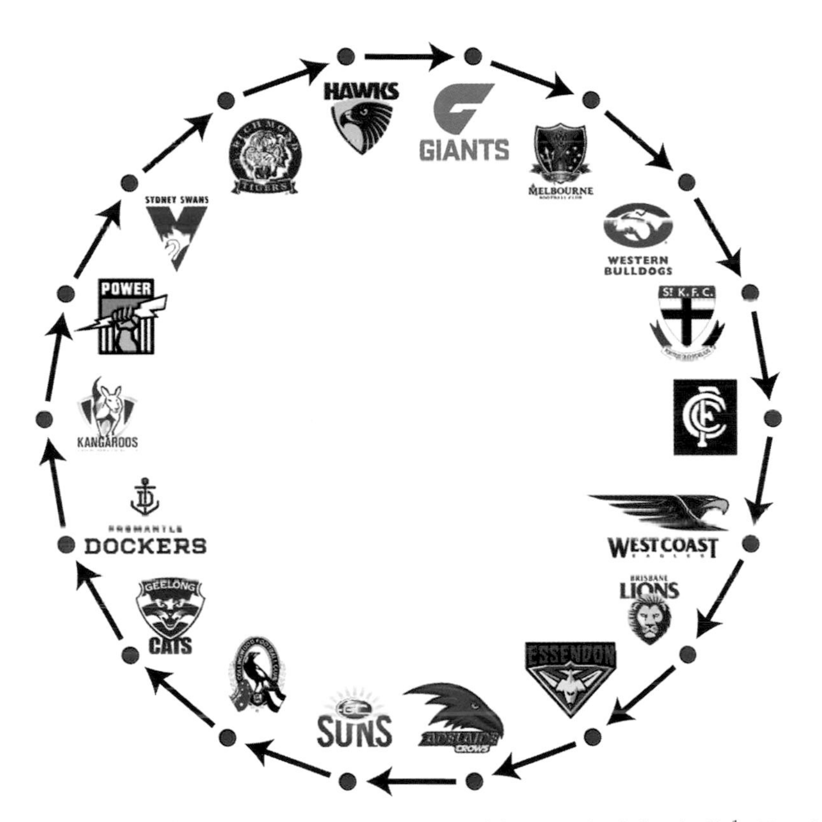

The weekend before last gave us a memorable round of football.[1] North Melbourne upset the Cats, and the Tigers thrashed Hawthorn. However, the truly momentous event was the win by Greater Western Sydney, thereby proving that GWS is the greatest team of all.

Say what? It took until Round 19 for Greater Western Sydney to record their first victory, and then they only beat Melbourne. That's akin to beating Heidelberg Reserves. So yes, GWS can be very happy that they scored a win, but what's the big deal?

Well, earlier in the season Melbourne beat Footscray, so clearly GWS are also a better team than Footscray. And, Footscray beat the Saints, who beat Carlton, and so on, right down to the "worst" team, Hawthorn. So, GWS must be the best team. Q.E.D.

[1] This column was written near the conclusion of the 2013 Australian (rules) Football League season.

OK, all that is pretty silly, of course. It's the desperate logic of wooden spooners.[2] (We Saints fan are masters at it.) However, there turns out to be some very pretty mathematics underlying the silliness.

Suppose we're following some competition, and we've reached the point in the season where all the teams have played each other.[3] It turns out that at this stage, and if we exclude the possibility of draws, then there must be at least one "best" team in the sense described above: Team A beat Team B, which beat Team C, right down to Team Z.

That may not seem so surprising, but in practise it can be tricky to find such a path through the teams. (It took us a while to locate the path through the AFL teams diagrammed above. The reader is invited to choose their favourite sport and give it a go.) It is even trickier to *prove* that there must always be such a path.

If we consider only three teams and their matches against each other, then there are two fundamentally distinct scenarios: either there is a team that has won both its matches or every team has had a win. In the former case, there is one clearly superior team, and in the latter case any team can be regarded as "best" in our peculiar sense.

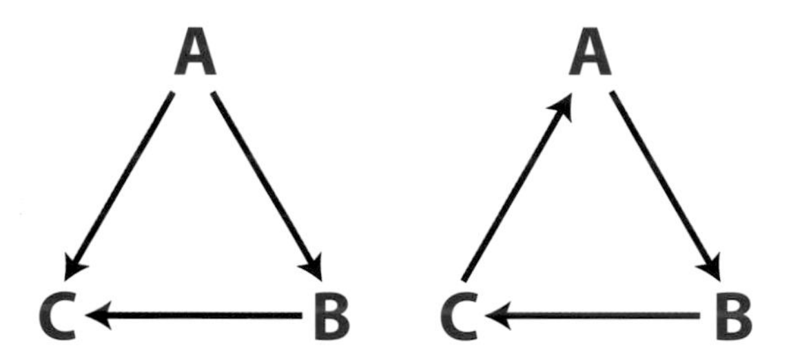

Suppose we now include a fourth team. We then have to consider how this team has fared against the three teams already considered.

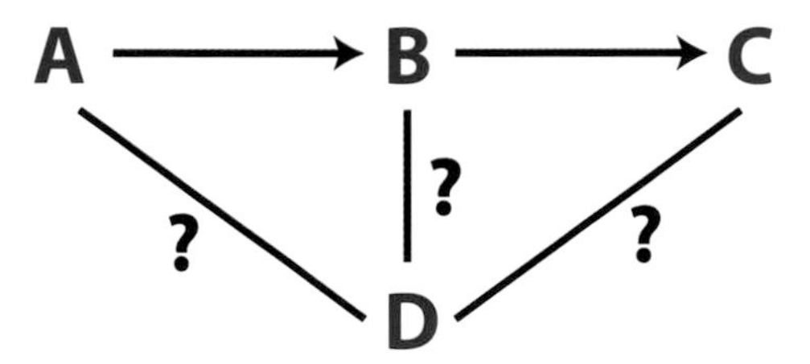

At this stage we can start drawing lots of diagrams and lots of arrows, but our task is simplified if we consider just three scenarios: First, Team D lost to Team C (in which case Team D gets included on the end, and Team A is still the best). Second, Team D beat Team A (in which case Team D gets included at the beginning as the new best team). Finally, Team D has beaten Team C and lost to Team A.

[2]The wooden spoon is the informal prize for the football team which wins the least number of games in a season.

[3]The Australian Football League schedule doesn't follow a simple procedure, or even a vaguely sensible one, but it was roughly at that stage at the time of writing.

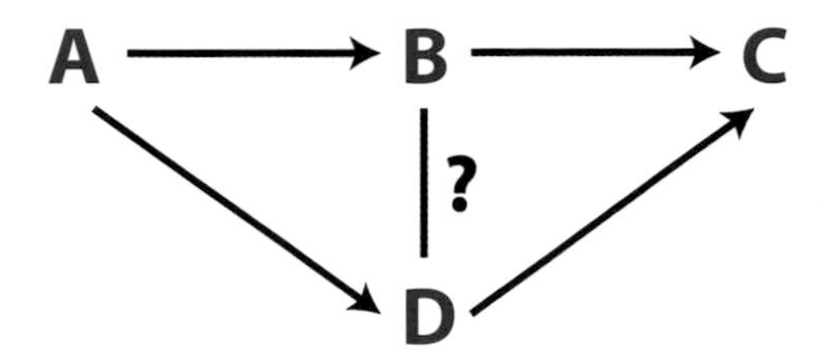

This last scenario requires a little thought, but the above illustration makes things clear. No matter the direction of the segment joining B and D, we'll have a path from A to C, with Team A maintaining its "best" status.

A competition between many teams can be analyzed in a similar manner. We incorporate the teams one by one, making a minor adjustment of the path at each stage to incorporate the new team.

That's a lovely argument; however, it only guarantees that there is *some* best team. There are definitely no promises that we can arrange for a struggler, such as GWS, to end up on top. Nonetheless, for many competitions that is the case: for *any* team we care to choose, there will be a path with our team as the "best". The Norwegian mathematician Øystein Ore dubbed this the *sportswriter's paradox*.

Let's consider Greater Western Sydney. What might suggest to us that there is a path that begins at GWS and makes its way through all the teams? Obviously, there was no hope until GWS won a match. Even then, if Melbourne had won no matches against a third team, we'd be stuck.

The trouble is that lowly teams tend to win against other lowly teams, if at all. Ore called such a group of teams *outclassed* if none of the teams in the group had won any matches except against other teams in the group.

Outclassed groups definitely occur. Before Round 19 GWS was an outclassed group all by itself, and only Melbourne's win over Footscray saved GWS and Melbourne from together forming an outclassed group. On the other hand, outclassed groups are less frequent than one might imagine. After all, it only takes one fluke win from any team in a group for the group to avoid being outclassed.

It is clearly impossible for a team in an outclassed group to be "best" in our path sense: no path can ever leave the group. However, it turns out that this is the only thing that can go wrong: if there are no outclassed groups, then any team can be arranged to be the best.

It's very nice mathematics, and it provides a little solace for the AFL cellar dwellers. However, there is a final mathematical touch to rub salt into the cellarly wounds.

If there are no outclassed groups, then, not only can we arrange for any team to be "best", we can arrange for all the teams to form a cycle. That is, the teams can be ordered so that Team A beat Team B, down to Team Z, which in turn beat Team A. (We completed the cycle in our AFL diagram above, with the Hawks beating GWS.)

So surprise, surprise—Greater Western Sydney are not so great after all. However, we'll not begrudge them their morale-boosting win, and we'll let them enjoy their "greatest" status for a week or so. Just as long as they don't score a couple more wins and relegate St Kilda to the basement.

Puzzle to ponder

Suppose a competition consists of just three teams. What does it mean then if no group is outclassed? Suppose now the competition consists of four teams. Try to convince yourself that if no group is outclassed, the teams can then be arranged in a cycle.

Part 3

Aussie heroes

With one very notable exception, Australia hasn't had much in the way of mathematical heroes. (Australia probably punches above its weight in producing good research mathematicians, but they don't typically make good subjects for popular writing.) So, the Australians we have written about are, by world standards, minor characters who somehow grabbed our attention. And, we have included one very impressive Aussie with absolutely no connection to mathematics.

CHAPTER 12

Around the world in 80 (plus 129) days

In 2010, after 209 straight days at sea, Jessica Watson completed her trip around the world, alone and unassisted. It would be a terrific achievement for anyone, and 16-year-old Jessica is now the youngest person to have accomplished it, eclipsing the 1998–1999 effort of then 18-year-old Jesse Martin. *Except* the respected website Sail-World.com claimed that Jessica didn't actually sail around the world!

Was Sail-World correct? In attempting to untangle this dispute, we had to consider what it actually means to sail around the world. It is a surprisingly tricky question.

Almost everyone would agree that following a (relatively) small circle around the Antarctic would be cheating. A precisely mathematical way to rule this out is to demand that the circle be as large as possible—the 21,600 nautical miles for the route around the equator. (A nautical mile is about 1.8 kilometers, slightly longer than a regular mile). However, this would require Jessica to drag her yacht over large stretches of dry land, which seems a trifle unfair. There are other maximum-size circles one could try, but no such circle passes entirely through the oceans.

Anyway, it is clearly fussy and impractical to demand such a precise route. A more natural requirement, which is often suggested, is that the route finish where it begins, and that along the way the route must pass through two points on exactly opposite sides of the Earth. Such points are called antipodal, with the most famous antipodal pair being the North and South Poles (of course, a useless pair for Jessica).

A practical difficulty with this requirement is that the natural sailing routes lie low in the southern hemisphere, and all the antipodal points are located in the northern hemisphere. So, in order to pass through antipodal points, a low southern route must make a significant detour into the northern hemisphere. This is exactly the type of route Jesse Martin followed.

A map of Jessica Watson's route makes clear that, though her yacht did poke its nose across the equator, it never passed through antipodal points. So, there is a sense in which it is fair to say that Jessica did not sail around the world.

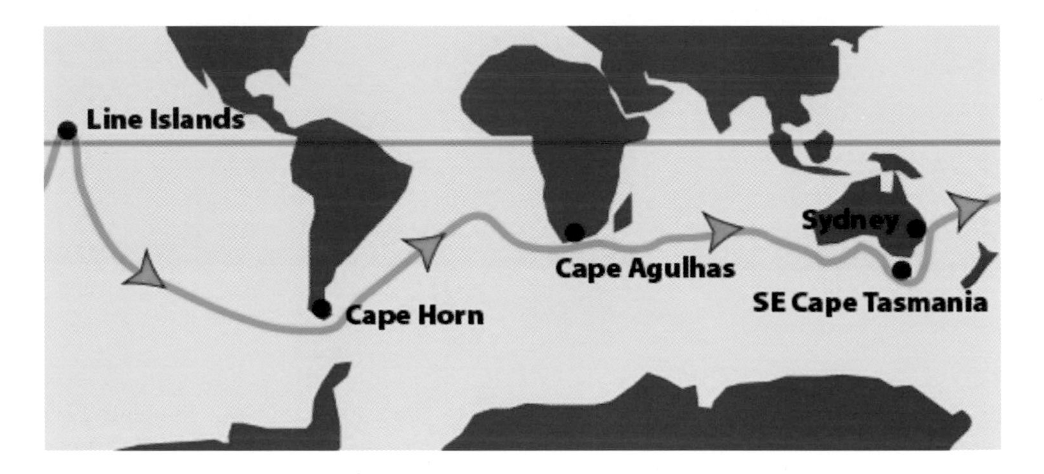

However, that is by no means the end of the story. In fact, the criticism from Sail-World has absolutely nothing to do with antipodal points.

Records for sailing "round the world" are governed by the World Speed Sailing Record Council, which has published a list of rules. To us, these rules seem quite arbitrary. For example, there is no requirement to pass through antipodal points, but the route is required to cross the equator (the reason for Jessica's northern detour). Why make people bother?

Nonetheless, the WSSRC can impose whatever rules they wish. The one rule of concern to Sail-World is *the shortest orthodromic track of the vessel must be at least 21,600 nautical miles in length.*

The 21,600 nautical miles is obviously there to ensure that the total distance travelled at least matches the length of an equatorial route. Fair enough. However, the *orthodromic track* is specifying how that distance must be calculated, and it is here that things get murky.

A path between two points is called orthodromic if it is the shortest possible path connecting the points. (Note that we're staying on the surface of the water here: submarines travelling along straight lines through the water are not permitted). But how does one use this to analyze Jessica's path?

To apply the distance rule, the reporters at Sail-World chose five land locations close to Jessica's route: Sydney, the Line Islands, Cape Horn, Cape Agulhas,

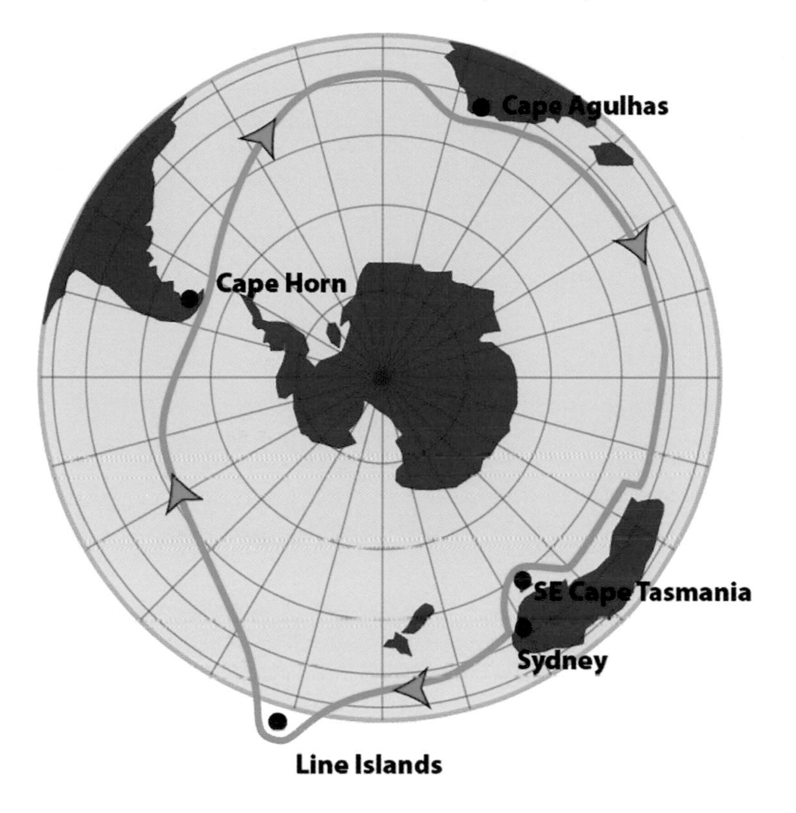

SE Cape Tasmania, and Sydney again to finish. The reporters then calculated the five orthodromic distances between adjacent points and summed these distances. They concluded that, even though Jessica's actual path exceeded 23,000 nautical miles in length, the orthodromic track of Jessica's route amounted to only 18,265 nautical miles, well short of the required distance.

We disagree. Even with the peculiar and confusing WSSRC rules at hand, we can make no sense of Sail-World's calculations. Firstly, the land locations seem arbitrary, since they only very roughly approximate Jessica's actual path. Secondly, we can see no reason to use only five locations, and using more would significantly increase the orthodromic track. In fact, using sufficient locations on Jessica's actual route would give an orthodromic track close to the 23,000 nautical miles that everybody agrees Jessica travelled. Finally, even if the WSSRC rules demand that the orthodromic calculation use land-based points (for heaven knows what reason), we believe that throwing in a few convenient islands near Jessica's route may raise her orthodromic track up to the required 21,600 nautical miles.

What if we're wrong, and Jessica's trip came up 3335 nautical miles short? As we read the WSSRC rules, Jessica could have made up the distance by sailing across Sydney Harbour a thousand or so times. Well, if that's what makes the difference between going "round the world" or not, who on Earth cares?

So, that's our conclusion. We believe the WSSRC rules are silly, and the complaints from Sail-World are sillier. To the extent that the requirements make any sense at all, we believe Jessica Watson has satisfied them. And, in the manner that most people seem to understand what going round the world means, we believe

Jessica has in fact done so. From the Maths Masters, and clearly from many others, congratulations!

Puzzle to ponder

We have suggested above that going around the world perhaps should require passing through a pair of antipodal points. Can you think of a silly route that goes through antipodal points but does not "really" go round the world? Can you think of a rule to exclude such silly routes?

CHAPTER 13

A prime mathematician

Prime numbers are maddening. Individually, they are as simple as possible. But, collectively, they are the most puzzling creatures in mathematics. Even for the brilliant Australian mathematician, Terry Tao.

The importance of primes is their role as numerical building blocks. For example, 60 equals the prime product $2 \times 2 \times 3 \times 5$. And, as Euclid proved, any whole number can similarly be factored into primes.

That is easy, but when we try to locate primes, things get weird. Yes, all primes excepting 2 are odd, but can we say anything else? The table highlights the primes below 180. Is there a pattern?

1	2	3	4	5	6	7	8	9	10	11	12	13	14	15	16	17	18	19	20	21	22	23	24	25	26	27	28	29	30
31	32	33	34	35	36	37	38	39	40	41	42	43	44	45	46	47	48	49	50	51	52	53	54	55	56	57	58	59	60
61	62	63	64	65	66	67	68	69	70	71	72	73	74	75	76	77	78	79	80	81	82	83	84	85	86	87	88	89	90
91	92	93	94	95	96	97	98	99	100	101	102	103	104	105	106	107	108	109	110	111	112	113	114	115	116	117	118	119	120
121	122	123	124	125	126	127	128	129	130	131	132	133	134	135	136	137	138	139	140	141	142	143	144	145	146	147	148	149	150
151	152	153	154	155	156	157	158	159	160	161	162	163	164	165	166	167	168	169	170	171	172	173	174	175	176	177	178	179	180

The full column $7, 37, \ldots, 157$ stands out. This is an *arithmetic sequence* of primes, where we're adding 30 each time. But, we've stopped our table just in time: adding 30 to 157 gives 187, and $187 = 11 \times 17$ is not prime.

The table suggests that the higher we look, the rarer primes become. Do they stop completely? If we have no pattern, how do we know there will always be a next prime, that there are infinitely many primes?

Once again, Euclid provided the answer. To make the question concrete, suppose we have stumbled upon the primes 2, 7, 13, and 23. How can we use these primes to guarantee there is another one?

Euclid's beautiful argument goes as follows. Collect the primes we already have, and form the number $N = (2 \times 7 \times 13 \times 23) + 1$. If N is prime, we are done, but that may not be the case. However, it is easy to see that N is not divisible by any of the primes we started with: dividing N by 2, 7, 13, or 23 leaves a remainder of 1. But, N definitely has prime factors, and so all those primes must be new. Quite. Easily. Done.

Notice that the number N will be big, and Euclid's process gives no easy method to actually identify the new primes. For example, in our case with $N = 4187$, what are the new primes? We'll leave that for you to puzzle over.

But didn't Euclid settle all these prime questions thousands of years ago? To demonstrate how troublesome primes still are, let us consider twin-primes. A twin-prime is a pair, such as 3-5 or 11-13, with a gap of 2 between.

It is easy to find examples of twin-primes. But, as primes become scarce, twin-primes become very scarce. Are there infinitely many twin-primes? No one knows!

Here is another problem. Suppose we look for arithmetic sequences of primes. Our table contains a sequence of six primes. Can you find an arithmetic sequence of seven primes? That's another (very difficult) puzzle for you.

Are there arithmetic sequences of primes of any length we like? Yes, but this was only proved in 2004 by Terry Tao together with the mathematician Ben Green. This led to Terry receiving the Fields Medal, the highest honor in mathematics. He was the first Australian ever to do so.

Can Terry Tao now solve the twin-prime problem? It is considered to be very, very difficult. But, Terry is very, very smart.[1]

Puzzles to ponder

1. What are the prime factors of 4187?
2. Find an arithmetic sequence of seven primes.

[1] We wrote this column in 2008. In 2013, the mathematical world was stunned by an announcement from mathematician Yitang Zhang. He had proved that there were infinitely many prime pairs that differed by at most 70 million—the first ever such theorem about twin-ish primes. Subsequently, Terry Tao and others have reduced that 70 million to 6. So, how hard can it now be to go from 6 to 2? Don't ask!

CHAPTER 14

Rubik's cube in ten seconds or less

In the beginning, nobody could solve The Cube. Not even Ernő Rubik, the cube's inventor.

Today, 43 years on, expert speedcubers can solve a scrambled cube in under 10 seconds. This includes Feliks Zemdegs, who became champion in 2010, when he was 16. Feliks, from Melbourne, Australia, set a new world record, solving five randomly scrambled cubes in an average time of 8.52 seconds.

Other people are not so quick. For lots, the clock is still ticking...

Most who have played with the cube can correctly arrange the top layer without assistance. Determining how to get further is not easy, though many people have solved the cube on their own. (Yes, your Maths Masters included.)

Millions of others have learned to solve the cube by following one of the ready-made recipes, available in books and on the internet. The key to these strategies is the use of specific combinations of cube twists. (By a twist we mean a rotation of one of the six faces of the cube, by either 90, 180, or 270 degrees). Each twist combination has the effect of moving a few targeted pieces, while leaving the previous work undisturbed. For example, one nice method begins by correctly arranging the top layer, and then uses just four twist combinations, over and over, to unscramble everything else.

These strategies are popular because they are not too difficult to follow. However, faced with a troublesome cube, they can easily require more than a hundred twists in total. Still, with practise, you should be able to reliably solve a cube in about a minute.

Feliks does not use such beginner's methods. The method Feliks uses requires memorizing more than a hundred separate combinations. With all these combinations at hand, Feliks can usually solve a cube with about 60 twists. Together with streamlined cube-handling, custom-built cubes, a good dose of talent, and LOTS of practise, Feliks can usually unscramble a cube in less than 10 seconds. But speedcubing is clearly not for the faint-hearted.

How much faster could Feliks become? Conceivably, that might depend upon him using even more combinations to lower the total number of moves required. So, how far is an average of 60 twists from being the best possible? Let's investigate.

There are 43,252,003,274,489,856,000 different scrambled states of the cube. How many of these take just one twist to solve?

Working backwards from a solved cube, we can choose to turn any one of six faces, and the chosen face can be twisted either 90, 180, or 270 degrees. That gives a total of $6 \times 3 = 18$ positions that can be solved with one twist.

Starting with one of these 18 positions, let's now make a second twist. There's no value in twisting the same face again, so we are left with $5 \times 3 = 15$ choices for this second twist. Together with the first twist, this shows that no more than $18 \times 15 = 270$ new positions of the cube can be solved with two twists. (It will actually be lower than 270 new solved positions, because some of the 270 two-twist combinations will simply be alternative methods for solving the same jumbled cube). Including a third twist, we find that that there are at most $18 \times 15 \times 15$ new positions that can solved with three twists. And so on.

Adding up these numbers, corresponding to combinations of up to 15 twists, we arrive at a monster number just below the total number of cube positions we gave above. This shows that there are definitely scrambled cubes requiring no fewer than 16 twists to unscramble.

In fact, using more sophisticated arguments and some serious number-crunching, it has recently been shown that any scrambled cube can be solved using at most 20 twists. Also, on average, 18 twists will suffice to unscramble a cube.

Comparing these numbers to Feliks's average of 60, there is seemingly plenty of room for improvement. But is that really true? Memorize zillions of combinations? Practise the instant recognition of a gazillion different positions, each requiring a different combination? Only time will tell what is really possible here, but shaving more than a second or two off of Feliks's record would be an incredible feat. Will the five-second barrier ever be broken?

Puzzle to ponder

How many twists does Felix perform per second?

CHAPTER 15

The wonderful Function of Michael Deakin

Younger readers may not previously have heard of Michael Deakin, but they should know of him. Mike lectured for many years at Monash University,[1] and many (older) math teachers would have been taught by him. More widely, Mike's decades of tireless work on the magazine *Function* has had a profound influence on thousands of Australian teachers and students, including a young Maths Master.

Function was a mathematics magazine published by Monash University until 2004. It was expressly intended for high school students, with the purpose of presenting articles "for entertainment and instruction, about mathematics and its history". *Function* was very ambitious, and it was very, very good.

Your Maths Masters have created a temporary online home for *Function* on our website, with all of the articles discussed below, and many more, available there.[2] Hopefully a more permanent home can be found in the near future.

Function first appeared in 1977, which happened to be the year a young Maths Master was in his final year of secondary school. On the advice of his wise teacher (a hello wave to Mrs. Baker), your Maths Master shelled out 90 cents for the first issue

[1] One of Australia's major universities, located in Melbourne.

[2] `www.qedcat.com/function`

of *Function*; he had never seen anything like it. *Function* was *about* mathematics: current research, historical roots, and playful problems and observations. That is not to deny the quality of senior secondary mathematics at the time, which was challenging and provided an excellent foundation for the university studies to follow. But *Function* was different. Every issue opened a new and unexpected window to the world of mathematics.

To give an example, *Function's* very first issue featured an article by one of our mathematical heroes, John Stillwell, on the four colour theorem. To understand what this famous theorem says, consider the following familiar map:

The problem is to colour the map so that no two states sharing a border are coloured the same. In our attempt above we've (somewhat carelessly) employed four colours to take care of all but one state, which would now require a fifth colour.

The four colour theorem states that we could have done better: the map above, and any map, can be coloured using just four colours. (In fact it is easy to see that three colours will suffice for our map above.) It is an enticing problem because it is so easy to understand and so, so difficult to prove. The theorem was only proved in 1976, and only with the aid of a computer to analyze over a thousand special cases. There still does not exist a computer-free proof.

Function provided an excellent article on what was then exciting and new research. It gave a clear description of the four colour problem together with an engaging historical overview. But even more engaging, the article included the essential details of the *five* colour theorem: for any map five colours will suffice. The argument wasn't easy going for a school student; however, no mathematical background was required, and with patience one could work through it.

Stillwell's article was a beautiful illustration of how difficult a simply stated question could be, and how a little insightful mathematics could (at least in part) conquer that difficulty.

Function was continually offering such wonderful articles, and, more than anyone, the sustained quality of *Function* was due to Mike Deakin. He was an editor for *Function's* entire existence, chief editor for most of that time, and he was far and away *Function's* most prolific contributor. Mike wrote 128 articles (as near as we could count them), cowrote five others and made Lord knows how many contributions as an anonymous editor. Many mathematicians, particularly Monash mathematicians, appreciated the value of *Function* and made very significant contributions. But Mike Deakin was *Function's* heart and soul.

It is impossible to convey the breadth of ground covered by Mike's articles. His first contribution, in *Function*'s second issue, was on catastrophe theory; it was important work on abrupt mathematical changes that became fashionable (and faddish) in the 1970s. It was a perfect topic for *Function*, a new and alive research area that could be explained just using polynomials. Mike's second contribution, in the very next issue, was a lovely article on magic hexagons, which turn out to be an astonishing variant of the more familiar magic squares.

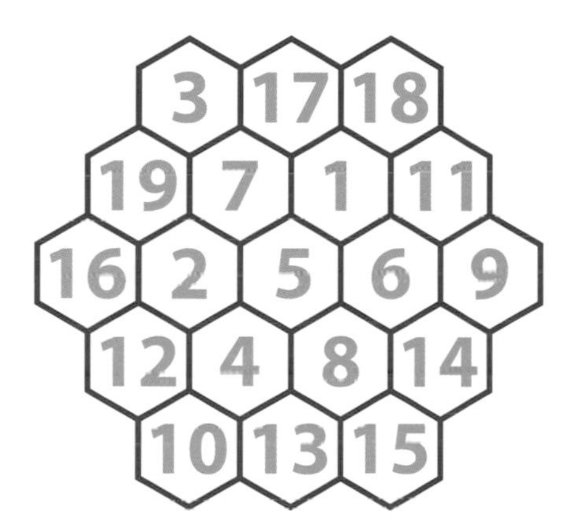

Above is an example of a magic hexagon, with the numbers 1 to 19 filling the cells, and with the numbers in each row (no matter the row's length or direction) summing to 38. The reader is invited to search for another magic hexagon (but may wish to first consult Mike's article for a hint).

Mike wrote often on the history of mathematics, including a fascinating article on Hypatia, generally considered to be the first woman mathematician. Mike carefully considered all the evidence to support such a claim and, somewhat disappointingly, concluded that Hypatia was probably not a mathematician of any stature, that she was likely "a transmitter of mathematics, not a creator of it". Hypatia was undoubtedly remarkable, "widely respected as a teacher—eminent, influential, even charismatic... But we have no evidence that she was anything more than this."

Mike also wrote very thoughtful articles on mathematical culture. He investigated the pre-prehistory of mathematics, the ancient remnants of mathematical thought that are evidenced in language. Mike wrote critically, and unapologetically, on the differences in mathematical cultures, on the distinction between "high numeracy" and "low numeracy" cultures. (Such concerns later led Michael Deakin to write a careful and scathing critique on an aspect of the Australian Curriculum, on the inflated role assigned to (essentially nonexistent) Aboriginal mathematics.)

Michael Deakin was a careful and critical scholar. However, Mike was also light on his mathematical feet, and he made many whimsical contributions. One of Mike's earliest contributions to *Function* was a short and clever argument that 2 + 2 = 3 (for small values of 2). In a similar spirit, Mike offered a mathematical argument that most people who have ever been born on Earth are still alive: so, at any time in the future, the odds are we'll be alive! Many more of Mike's contributions were written in a similar spirit.

Writing about Michael Deakin and his incredible output of clear and engaging mathematics, one's thoughts invariably turn to the brilliant Martin Gardner. As a mathematical populariser no one can compare to Gardner, however Mike Deakin is as close as Australia has ever produced. And more than a populariser, Mike was an active mathematician and historian. His many interesting and original contributions appeared in the *American Mathematical Monthly*, the *Gazette of the Australian Mathematical Society*, and in many other journals.

However, it is Mike's inspiration to young mathematicians, and in particular his amazing work on *Function* that will be Mike Deakin's enduring legacy. In 2003, Mike received a prestigious and much-deserved award from the Australian Mathematics Trust, honoring his great work. He is missed, and he is irreplaceable.[3]

Puzzle to ponder

Colour a map of the continental United States with four colours. Can it be done with just three colours?

[3]Our friend and colleague Mike Deakin died in August 2014. We wrote this column as a tribute to him.

CHAPTER 16

A Webb of intrigue

Meet Robert Webb, one of Melbourne's mathematical stars (even if most of Melbourne is unaware). He is a professional programmer and the author of *Stella*, the award-winning program which is a tool for visualizing and creating polyhedra. The above photo features Robert posing with some of his beautiful, handcrafted models.

Robert is a wonderful illustration of what can happen when a teacher introduces some beautiful mathematics to schoolchildren. Many of our readers will be familiar with the Platonic solids, shown below. These are the five polyhedra that can be constructed using just one type of regular polygon, with every corner of the polyhedron identical.

Robert still remembers building models of the Platonic solids as part of a school project and being intrigued by their beauty. It started a life-long obsession with these shapes and the vast families of their geometrical cousins.

Robert started working on *Stella* in 1999, when he was primarily interested in generating uniform polyhedra. This family includes the Platonic solids, the

Archimedean solids (similar to Platonic solids, but where more than one type of regular polygon is used), and many more. Then, *Stella* learned how to "stellate" polyhedra, dramatically increasing the range of objects that the program could handle. Still later, Robert added the ability to create and print nets, enabling him to build precise cardboard models of very complex geometric shapes.

Stella has a range of incarnations. Simplest is *Small Stella*, which has 300 built-in models that can be viewed in 3D, animated and manipulated in a myriad of ways; it includes the Platonic and Archimedean solids, and geodesic spheres, to name just a few. *Great Stella* is more advanced and includes a much larger range of built-in models as well as the tools for building zillions of new ones. *Stella4D*, which your Maths Masters have been using for years, is the deluxe version. It includes everything in *Great Stella*, together with support for weird 4-dimensional polythings. *Stella4D* enables one to obtain beautiful and insightful images of fabled 4-dimensional creatures such as the hypercube and the 120-cell. Finally, Robert also released *MoStella* an Android app version.

To give a taste of what's involved in creating one of Robert's models, let's take a look at the "multicube", five interlaced cubes as illustrated in *Stella* (left) together with Robert's paper model (right). How did Robert do it?

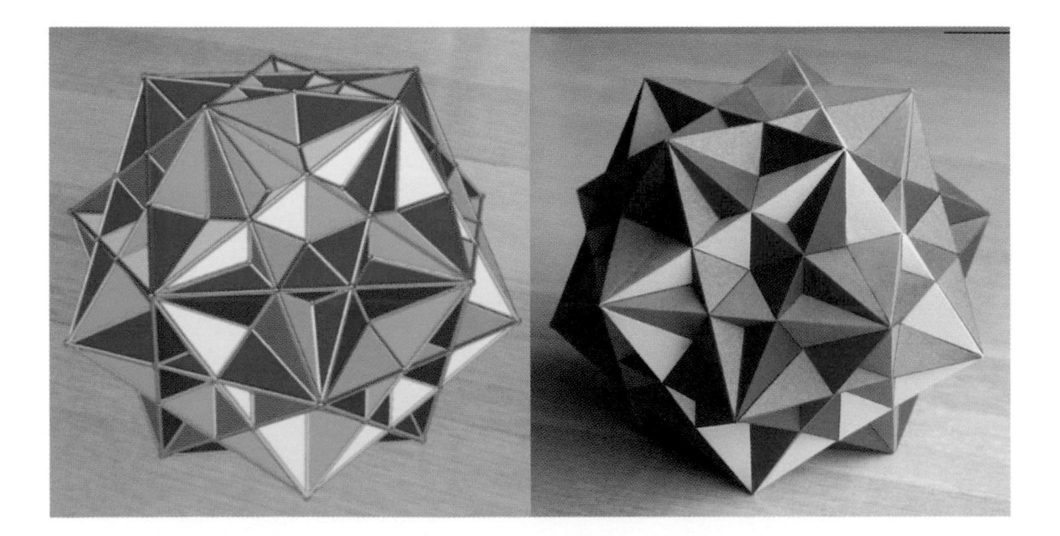

Every child will at some point get to make a cube from paper, usually beginning with a flat net of the cube:

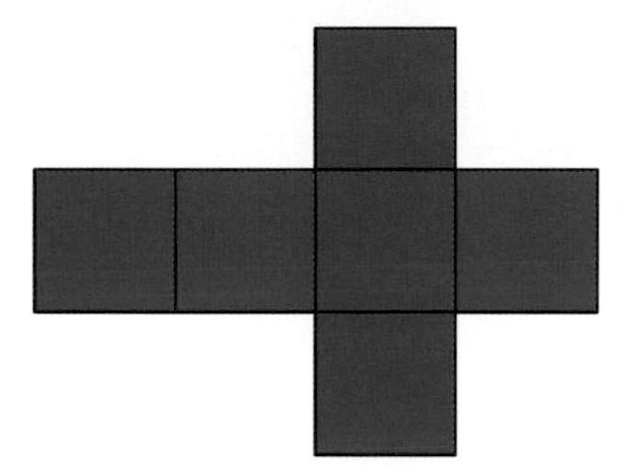

Similarly, to make our crazy multicube, we might simply ask *Stella* to create the appropriate net. We then grab a convenient child, give her the net and some sticky tape. Voilà! There's your shape.

Well, maybe not. Handling a net of 360 triangles is pretty much impossible for anyone, child or not. And, there may be a more fundamental problem: it is not even clear that the multicube has a net.

It may come as a surprise that not all polyhedra have nets. A simple example is the witch's hat pictured below, consisting of three triangles and three trapezia.

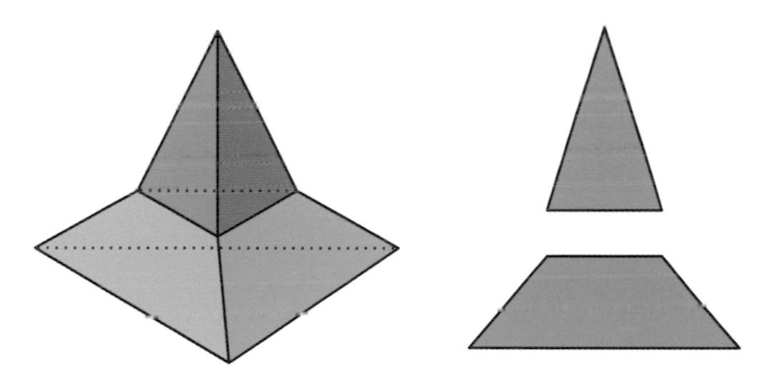

What goes wrong? If a polyhedron has a net, then it can be obtained by cutting one edge at a time until whatever is left is still connected but folds flat without overlap. However, any such attempt with the witch's hat results in overlaps. Two failed attempts are pictured below.

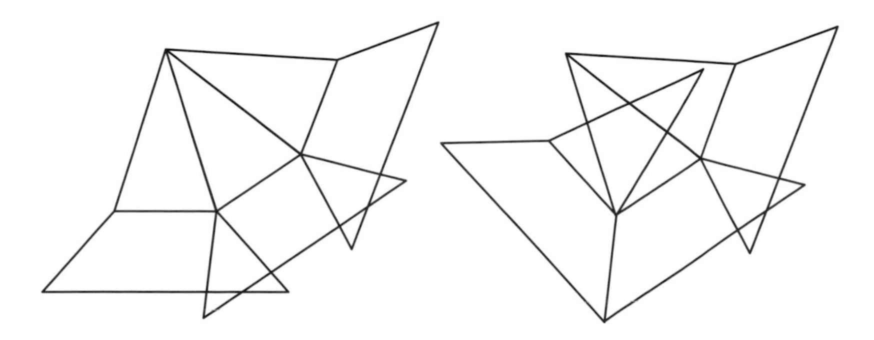

There is much more to nets than meets the eye, and surprisingly little is known about them. It is not even known whether all convex polyhedra have nets. So, does the multicube have a net? We're not sure. In any case, a net for any of Robert's complex creations would be much too unwieldy to be of practical use.

Instead, Robert creates a number of cunning subnets, each destined to be a carefully chosen part of the whole polyhedron. Robert's clever and careful step-by-step guide for constructing the multicube can be viewed at his website.[1]

Puzzle to ponder

Given six identical squares, how many different nets of the cube can you make with these squares? (Two nets are counted as different if they are not congruent as shapes in the plane. That is, one net cannot be moved and/or flipped so as to be placed identically on top of the other.)

[1] www.software3d.com/5Cube.php

Part 4

Melbourne, City of Mathematics

We have no idea why, but for some reason our hometown of Melbourne has a wealth of mathematically inspired architecture (plus, one tree with a proud mathematical pedigree). We could have written a whole book on this one topic, but we content ourselves here with presenting some of the highlights.

CHAPTER 17

A very peculiar Storey

One of your Maths Masters once received an invitation to attend a function at RMIT's Storey Hall on Swanston Street.[1] The fancy invitation card included a photo of Storey Hall's strikingly tiled walls, together with some mathematical background:

> The dynamic tiles featured throughout the venue were inspired by renowned Oxford professor of mathematics, Roger Penrose. The venue reflects his discovery that instead of the 20, 426 types of geometric shapes required to cover a continuous surface in a nonperiodic pattern, it could be done with just two: a rhombus of 54 degrees and another of 72.

Unfortunately, this description, courtesy of the Victorian Department of Education (VDE), is hopelessly garbled and almost totally wrong. *Plus ça change...*

[1] The Royal Melbourne Institute of Technology, on one of the main streets in the centre of Melbourne.

Storey Hall is one of Melbourne's architectural icons. So, it seems worthwhile to clarify exactly what is mathematically special about its famous façade.

Take some flat shapes, triangles or squares or what have you. We say that these shapes *tile the plane* if it is possible to cover the whole plane with copies of the shapes, without overlaps (except on the edges) and without gaps.

It will probably come as no surprise that the plane can be tiled with equilateral triangles, and also with squares and with regular hexagons.

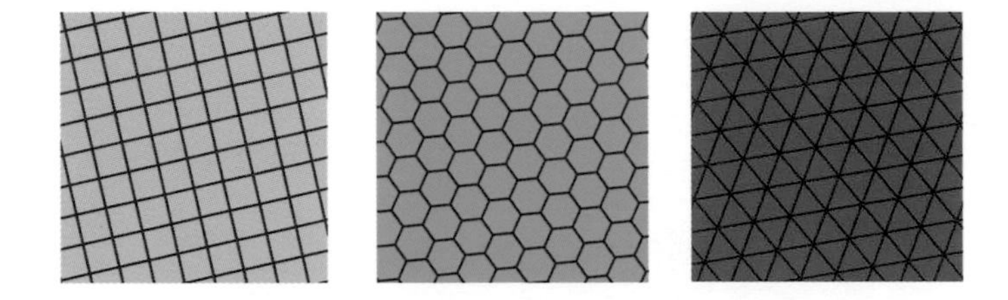

However, this is just the beginning. Staying with regular shapes, we can tile the plane with combinations, as in the following examples.

The above tilings are all *periodic*. That is, we can superimpose a parallelogram grid on the tiling so that the inside of each parallelogram looks the same, as pictured below. (Actually, there are subtleties with the precise meaning of "periodic", but the parallelogram idea suffices for our discussion here.)

The focus of this chapter might suggest that periodic tilings are the boring ones. However, one should not ignore the stunning periodic possibilities, as exhibited, for example, in Islamic art and in the work of M. C. Escher.[2]

Anyway, what of nonperiodic tilings, which the department's invitation suggests are a big deal? In particular, the implication is that a nonperiodic tiling requires at least two different types of tiles. That is not true, and indeed an ordinary square tile suffices: we simply have to shift one row of the previous, periodic tiling.

[2]See Chapter 1.

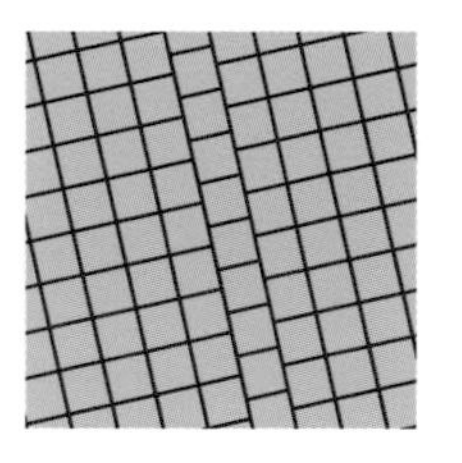

Pretty boring. Below are two much prettier nonperiodic tilings, based upon a single isosceles triangle. And, for another striking example, we direct you to the walls of Federation Square, which we discuss in the next two chapters.

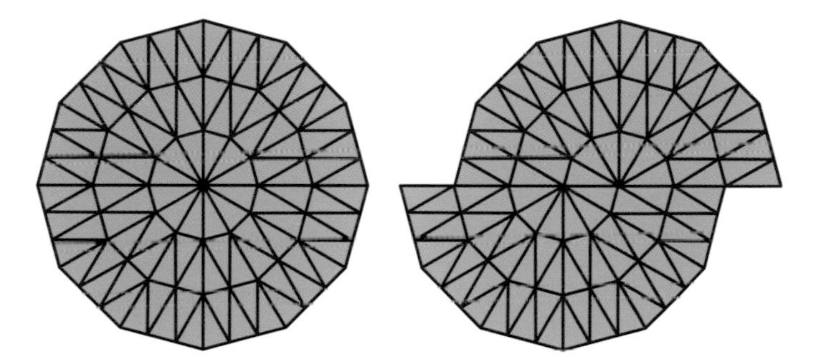

Clearly, nonperiodic tilings are easy to come by. So, what's the big deal?

Notice that if a collection of tiles can be used to tile the plane, then there seem to be many different ways to do the tiling. For example, the isosceles triangle easily gives a third, periodic tiling:

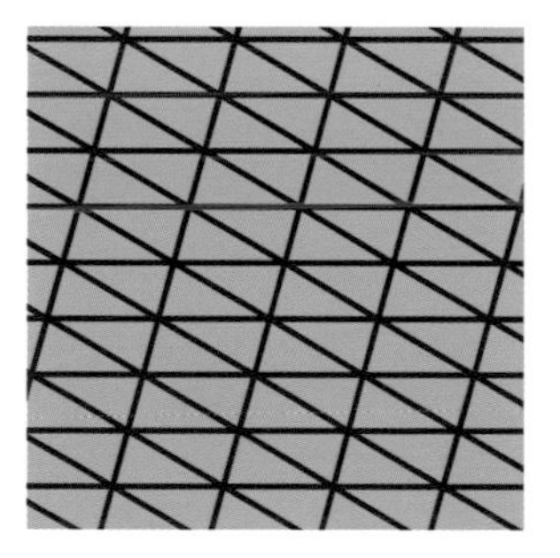

That brings forth the real question: If we have a nonperiodic tiling, can the same tiles also be used to construct a periodic tiling? If the answer is "no", if the collection of tiles can *only* be used to tile nonperiodically, then the collection is called *aperiodic*.

For a long time, no one knew whether there existed an aperiodic collection of tiles, and in fact it was suspected that there did not. Then, in 1964, the mathematician Robert Berger constructed an aperiodic collection of 20,426 tiles—the magic number referred to in the VDE's invitation.

Berger and others went on to find much smaller collections of aperiodic tiles. In 1973, Roger Penrose found an aperiodic collection consisting of six tiles, and soon after he discovered an aperiodic collection consisting of just two tiles.

Penrose came up with a number of different (though closely related) aperiodic pairs of tiles. One particularly nice pair consists of two rhombuses, embellished with little jigsaw cuts and knobs.

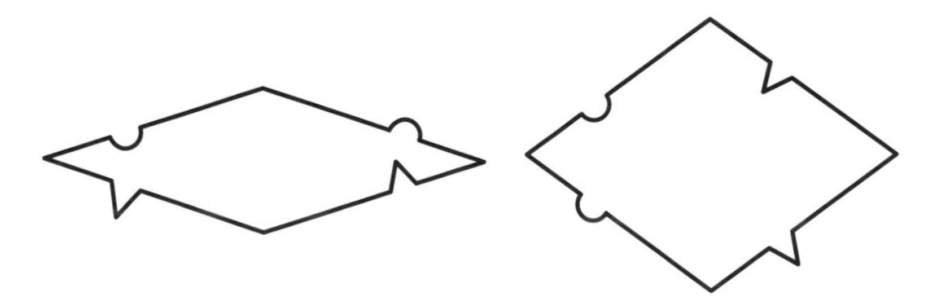

The point of the jigsaw business is to restrict the ways in which rhombus tiles can be pieced together. If the cuts weren't there, we could easily make a periodic tiling, just as we did with the isosceles triangle, and that's exactly what we hope to make impossible.

An attractive alternative to making jigsaw cuts is to add two colours to the tiles:

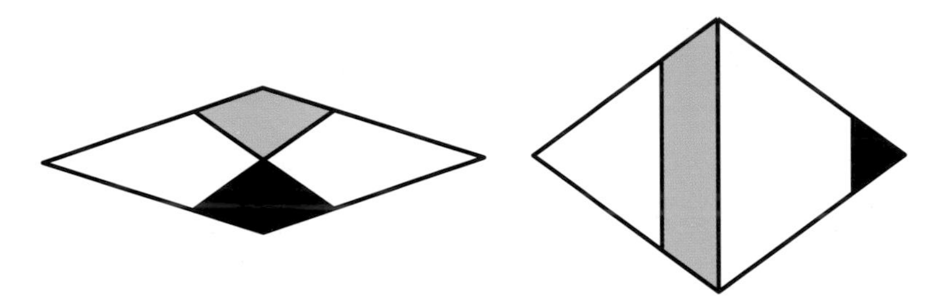

We then require that in any tiling, the colours must match across the edges. It is a variation on these coloured rhombuses that decorate Storey Hall.

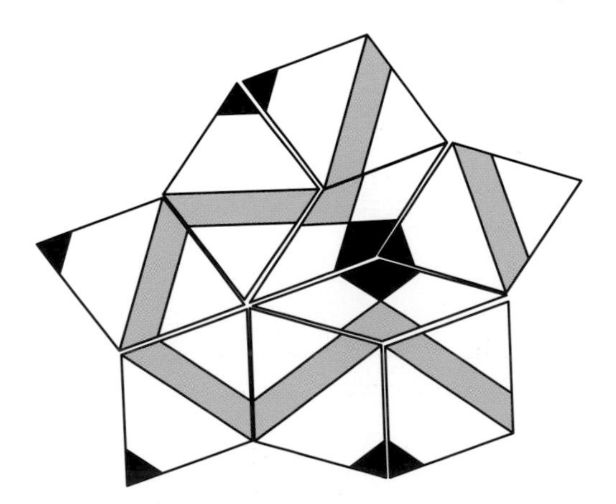

There is plenty that is mysterious about the Penrose tiles. For example, how did Penrose tiles lead to the 2011 Nobel prize in chemistry?

More fundamentally, how do we *know* that Penrose's rhombuses can be used to tile the whole plane? And, how do we *know* that they cannot be used to tile periodically? These are subtle questions and too deep to go into here: the inquisitive reader is referred to the excellent articles by David Austin.[3]

Let's now take a careful look at the Storey Hall tiles:

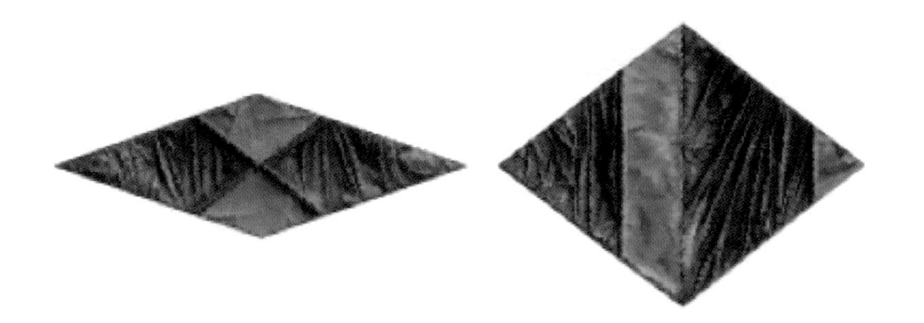

They are rhombuses of the same proportions and the light-coloured regions are in the same locations as the regions on the Penrose tiles. However, the matching regions on the Storey tiles are all one colour rather than two. This is a very important difference.

With only one colour to be matched, many more tilings are now possible. And, as is the case for the example below, some of these new tilings are periodic.

We reiterate—what is mathematically special about the original Penrose tiles is that the only possible tilings are nonperiodic. Now, the reduction from two matching colours to one has completely destroyed this property.

Why do it? Who knows?

Still, the Storey Hall tiling is very attractive, and perhaps all is not lost. One can take the tiling and attempt to redo it with proper Penrose tiles. It almost works.

---

[3] www.ams.org/samplings/feature-column/fcarc-ribbons

Below is a reconstruction of the top part of Storey Hall's façade with proper Penrose tiles. It works, as far as it goes.

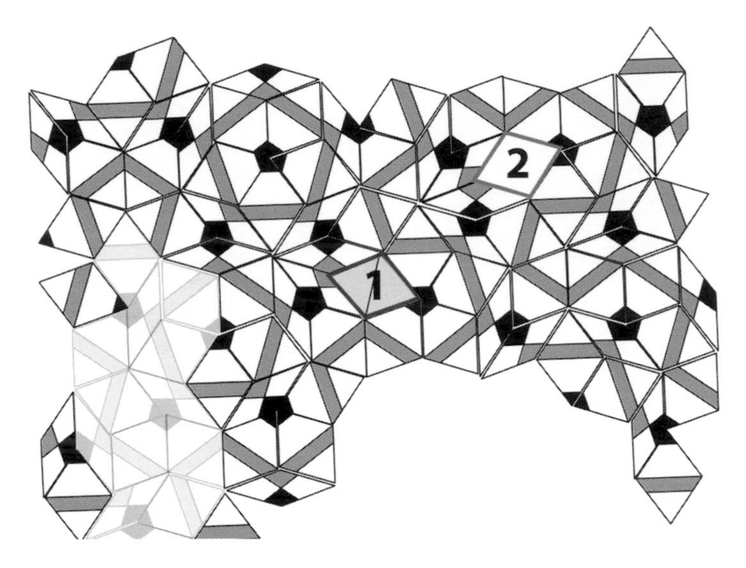

Notice that the façade is missing tiles at spots 1 and 2. This turns out to be unavoidable. Though the gaps are just the right size to be filled with the fatter rhombuses, this cannot be done so that the colours match, even with Storey Hall's single-colour tiles.

This highlights a central and perplexing difficulty of Penrose tiles: even though we have told you that the whole plane can be tiled, it is not at all clear how to actually do it. Simply laying down Penrose tiles one by one and hoping for the best will likely fail: a configuration with unfillable gaps can easily arise. (For the much fuller story of Penrose tiling, we again refer you to the articles by David Austin.)

In particular, as we noted above, the tilings on the Storey Hall façade cannot be extended to fill the gaps at spots 1 and 2. However that does not appear to be an architectural goof. Rather, the gaps seem to have been arranged deliberately, to highlight the peculiar manner in which Penrose tiles work.

However, there are other, seemingly less deliberate, problems. Below is a close-up of the lower left part of the façade.

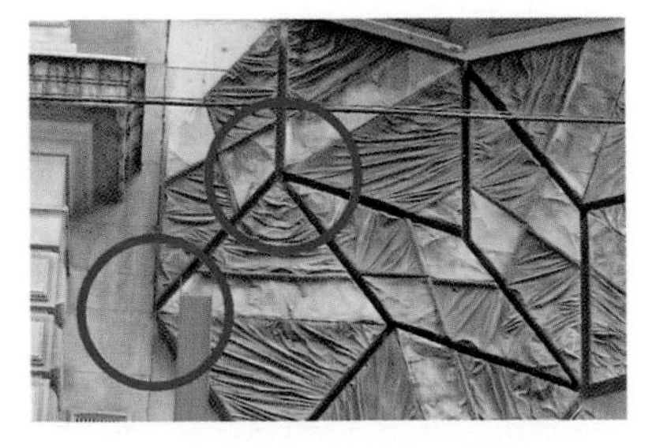

The matching rules, even Storey Hall's simplified one-colour rules, are violated at the two corners circled. Perhaps the architects would argue some aesthetic purpose for this as well. However, to us these corners appear to be straightforward goofs. (Similar problem corners also occur in the tiling inside Storey Hall).

There are other ways in which the Storey Hall tiling is puzzling, though none as glaring as the above issues. Anyway, if you get the chance to visit Melbourne, maybe stop to peruse Storey Hall, and see what does and does not make sense to you.

Notwithstanding all our criticism, we do really like Storey Hall. It is a beautifully designed, very functional building with stunning walls and façade. We simply wish the architects had considered more carefully the nature of Penrose tiles. Or, to the extent the "mis-tiling" was done deliberately, we wish we could understand why.

Puzzle to ponder

1. Can you find a mistake in the part of the wall on the inside of Storey Hall pictured below?

2. Find another mistake in the VDE's description of Penrose tiles.

CHAPTER 18

Federation forensics

Looking at the beautiful tiling of RMIT's Storey Hall,[1] we were surprised by some of the tiling's mathematical oddities, and that got us to pondering, What about Melbourne's other famous geometric façade?

The *pinwheel tiling* of Federation Square begins with a right-angled triangle of just the right shape. We then add four identical triangles to create a super-triangle of the same shape:

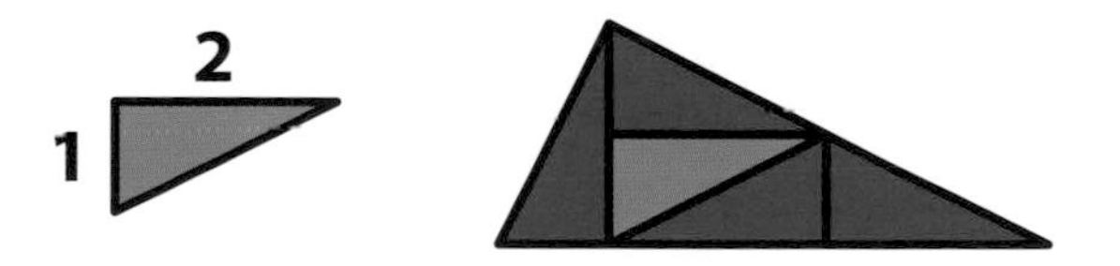

[1] See Chapter 17.

Next, in exactly the same way, we add four super-triangles to build a super-super-triangle:

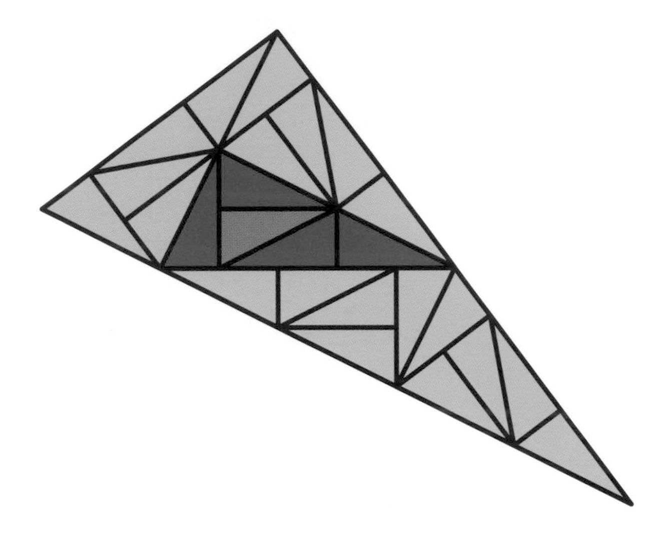

Since each new triangle is the same special shape, we can repeat the process indefinitely, filling the whole plane with successively super-sized triangles. The result is definitely reminiscent of Federation Square:

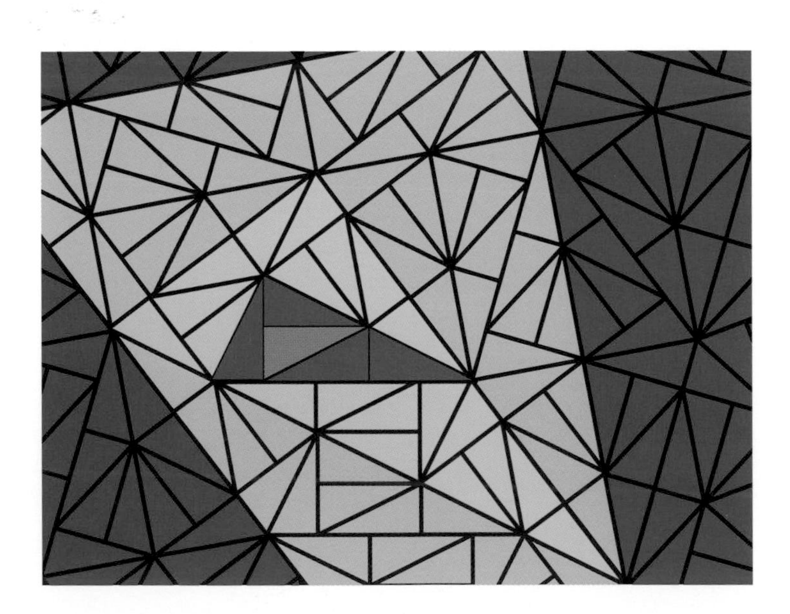

We'll now take a closer look at Federation Square, beginning with the ACMI wall, facing Flinders Street. If the tiling on this wall is really part of the pinwheel tiling, then we should be able to group the triangular tiles into super-triangles.

This is indeed the case: the orange triangles outlined below are all super-triangles, each consisting of five basic tiles.

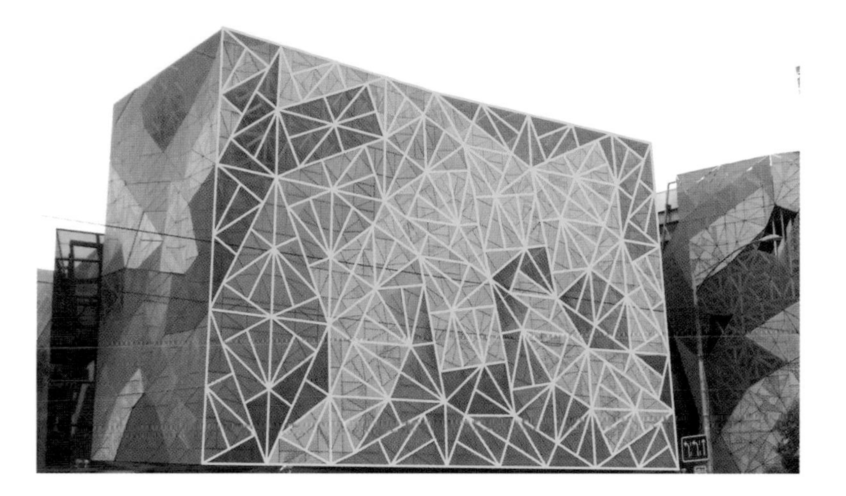

In turn, the orange super-triangles can be combined into green super-super-triangles:

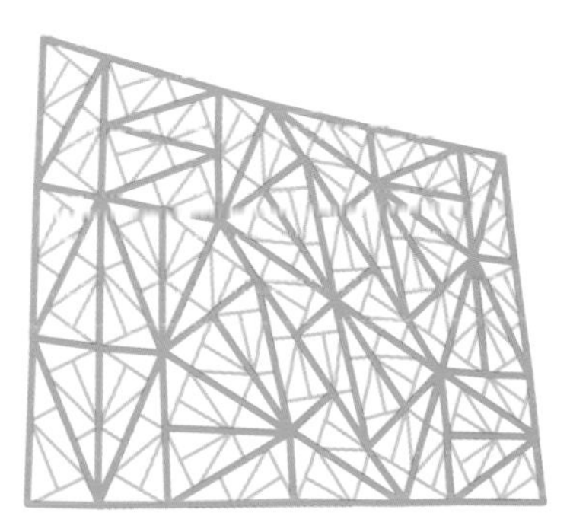

However, at the next stage we run into trouble. It partly works, as the purple triangles outlined below are the desired super-super-super-triangles:

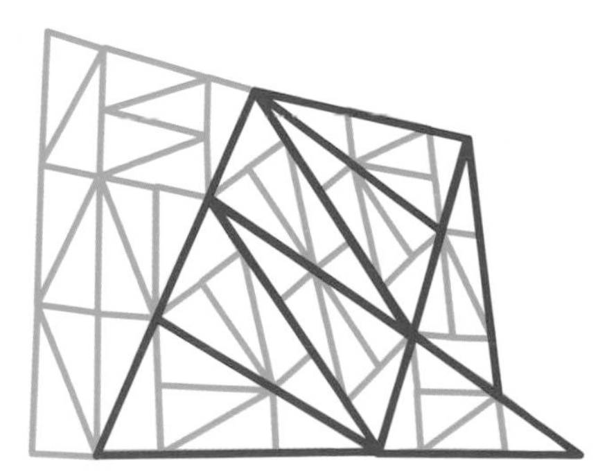

Alas, the green tiles on the left side of the wall cannot be combined in this manner. Close, but no cigar.

Now, was this carelessness on the architects' part, or were they forced to cheat a little? We suspect the latter.

In fact, it is difficult to find rectangular portions of the pinwheel tiling of just the right size. Taking the short side of a basic triangle as our unit, the ACMI wall is 45×30 units. We can now hunt for rectangles with these proportions.

Rectangles within the pinwheel tiling are of two types: either the perimeter consists entirely of hypotenuses of the basic tiles, or the perimeter contains no hypotenuses. The green and orange rectangles outlined below are examples of the two types.

For the Federation Square walls, the architects chose to go with no-hypotenuse rectangles. This means that the ACMI wall could have been modeled on the 6×4 orange rectangle. Of course, the tiles required to do so would have been massive.

However, we can also consider each triangle in the previous diagram to be a super-super-triangle. In this case the orange rectangle would have dimensions 30×20. Or, going to the next possible level, we can regard the orange rectangle as being 150×100.

But is there actually a rectangle of size 45×30, or even close? We don't know. So, though the architects could definitely have made the ACMI wall an exact pinwheel tiling, this may well have necessitated basic tiles of a significantly different size.

Moreover, so far we have only considered one wall: even if tiles were chosen of a size ensuring one wall could be tiled perfectly, those tiles may well fail to work for the next wall. In fact, this seems to have been what happened.

Consider the northern wall of Federation Square, pictured below. It is 20×30 units, which means that it definitely can be properly tiled, and indeed it has been. The tiling is exactly the one we indicated above; each orange triangle outlined above is a super-super-triangle of the pinwheel tiling.

We investigated three other walls at Federation Square. The smallest wall followed the pinwheel tiling exactly, and the other two were similar to the ACMI wall: they followed the pinwheel rules up to the super-super-triangle level, but then things broke down.

Anyway, we now have a clear sense of what does and what does not work in Federation Square. Overall, we're impressed. True, it would have been wonderful

to have all the walls follow the pinwheel rules exactly, but to complain would be churlish; the architects have done a beautiful job with a very unwieldy mathematical construct.

Puzzles to ponder

1. Below is one of the walls of Federation Square. The tiles are all right-angled triangles, with shorter sides 60 centimeters and longer sides 120 centimeters. How many tiles have been used to cover the wall?

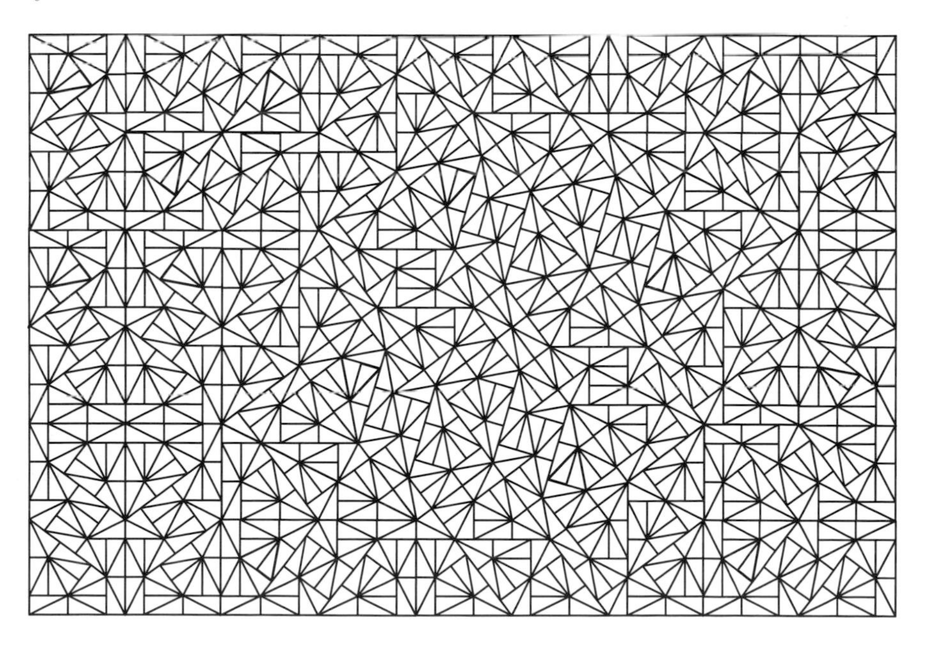

2. How many rectangles composed of triangles can you find in the following patch of the pinwheel tiling?

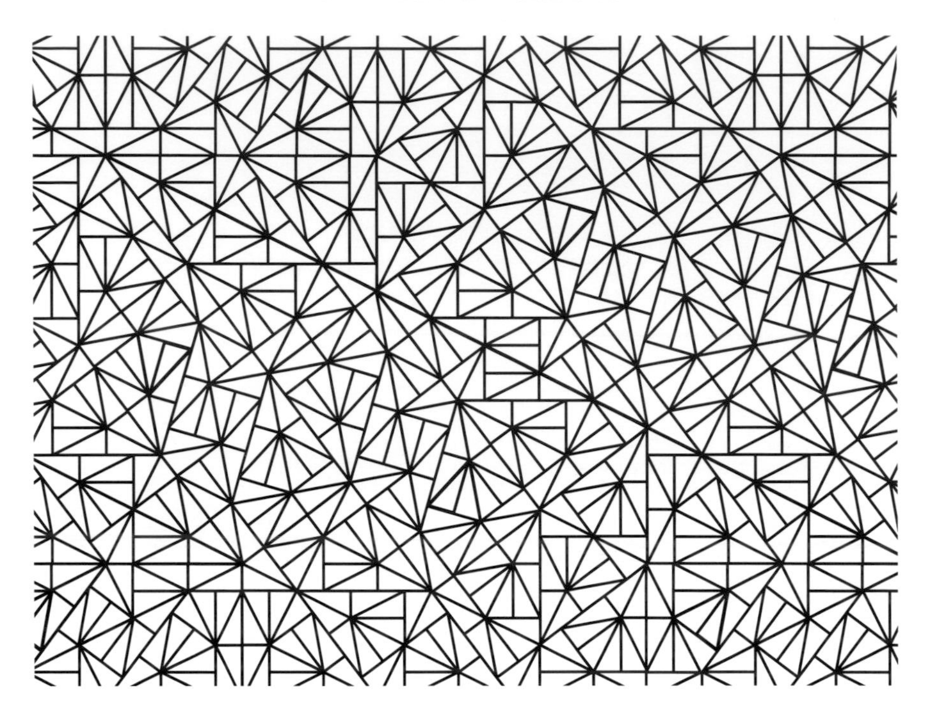

CHAPTER 19

More forensicking

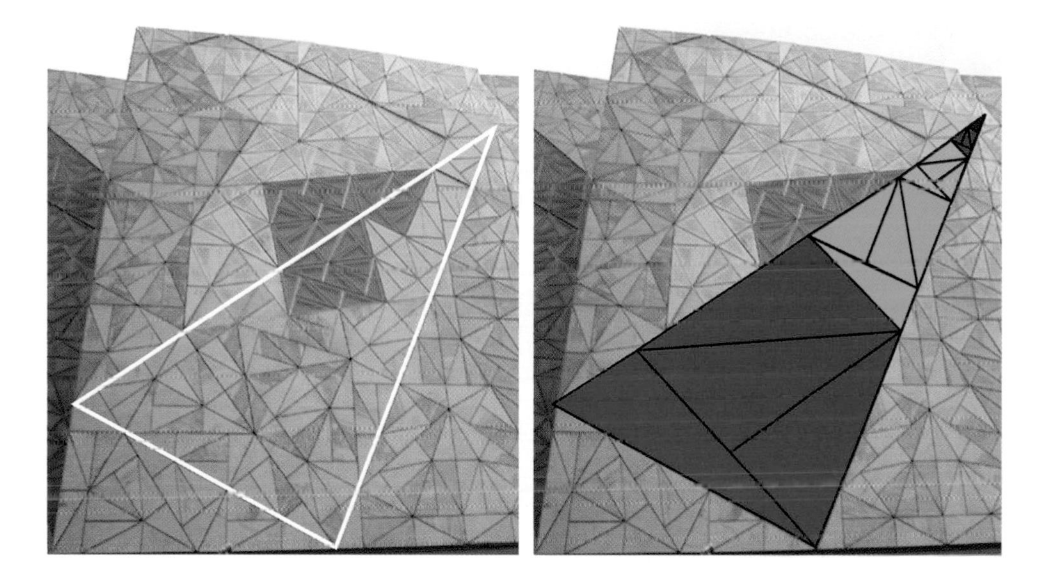

So, here we are, back in Federation Square. We've already taken a very close look at the square's triangular tiling, to see what the architects got right and (arguably) what they got wrong. However, we were inspired to write more by an unexpected encounter with Federation Square's doppelgänger. It appeared in an article by mathematician Roger Nelson,[1] author of the marvellous book *Proofs without Words* and its sequel.[2]

To explain this appearance, let's first recall the special property of Federation Square's façade. We begin with a big, judiciously chosen portion of the façade, as outlined in the picture above.

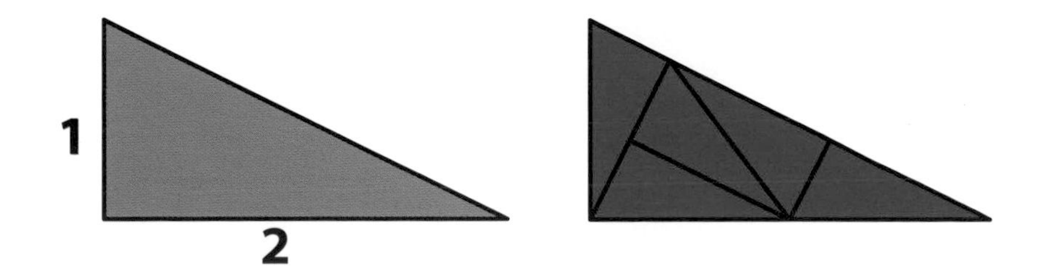

[1] *Proof without words: Right triangles and geometric series*, Mathematics Magazine, **79** (2006), 60.

[2] Mathematical Association of America, 1993 and 2001.

This triangular piece is just the right shape so that it can be replaced by five red sub-triangles of the same shape. Then those sub-triangles can be replaced by green sub-sub-triangles, and so on.

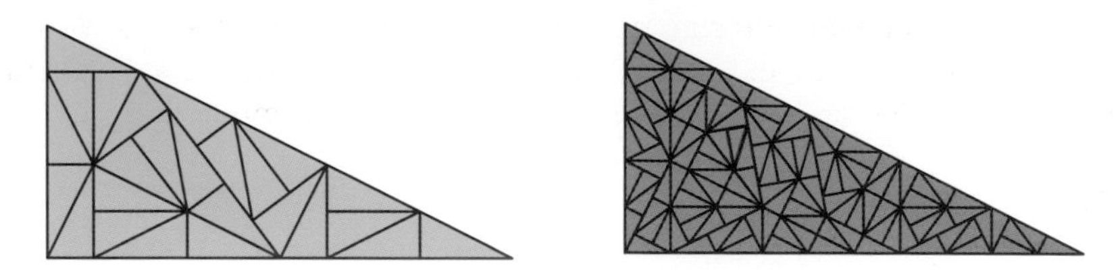

We can stop with the orange tiles pictured, giving us part of the Federation Square tiling. However this is when the doppelgänger appears, enticing us to keep going.

Let's begin again with the five red sub-triangles. However, at the next stage we only replace the rightmost red triangle.

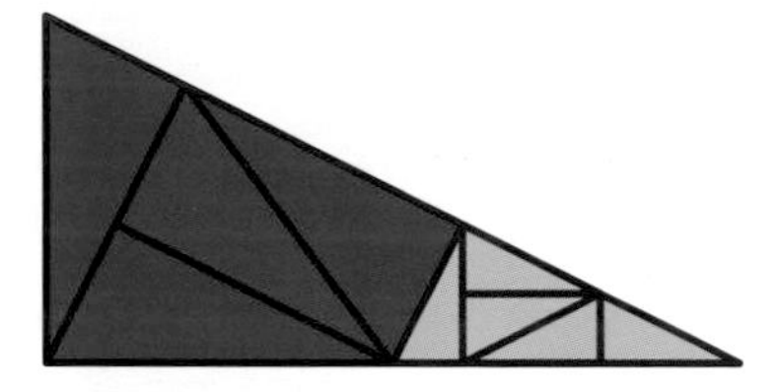

Similarly, at the next stage we only replace the rightmost green triangle.

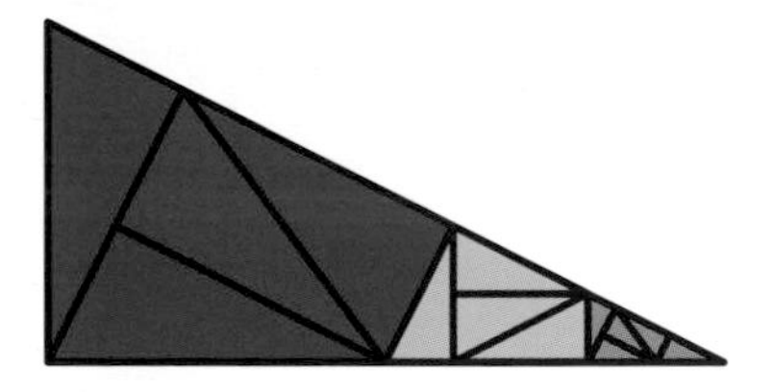

We do the same at the next stage, and the next, and... we just don't stop. There are infinitely many stages, where at each stage the rightmost minuscule triangle is cut into five micro-minuscule triangles.

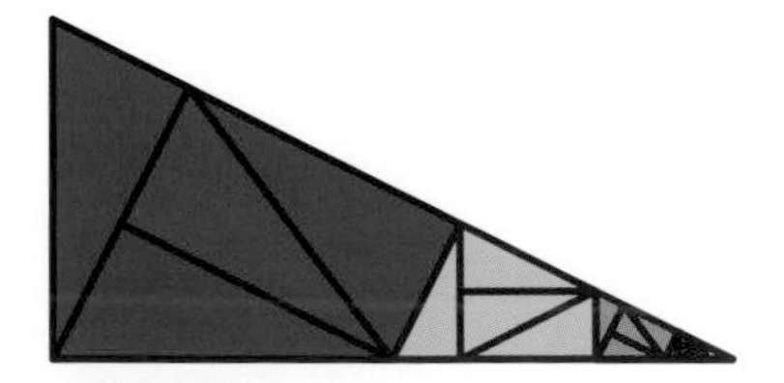

It's a cool picture but is there a point to it?

Recalling the mantra "one-half base times height", we see that the original brown triangle has area 1. It follows that each of the four red triangles have area $1/5$, the green ones have area $1/5 \times 1/5$, the orange ones have area $(1/5)^3$, and so on. Summing the areas of all the sub-triangles, our diagram shows at a glance that

$$\frac{4}{5} + \frac{4}{5^2} + \frac{4}{5^3} + \cdots = 1.$$

Or, after dividing both sides by 4,

$$\frac{1}{5} + \frac{1}{5^2} + \frac{1}{5^3} + \cdots = \frac{1}{4}.$$

That is a simple and beautiful summing of an infinite geometric series. Not surprisingly, other infinite series can be similarly summed, and Roger Nelson uses different right-angled triangles to calculate four such sums. For example, the diagram

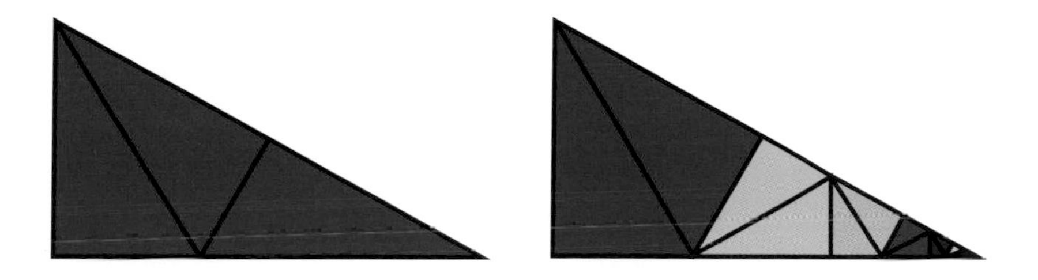

demonstrates that

$$\frac{1}{3} + \frac{1}{3^2} + \frac{1}{3^3} + \cdots = \frac{1}{2}.$$

You can probably guess the general formula:

$$\frac{1}{N} + \frac{1}{N^2} + \frac{1}{N^3} + \cdots = \frac{1}{N-1}.$$

This formula is true for any number $N > 1$ (and any $N < -1$), and there is a pretty and well-known algebraic proof of the formula. However, we'll stick to geometry and to positive whole numbers.

We have taken care of $N = 3$ and $N = 5$, so can we find triangles to demonstrate the formula for other whole numbers N? It is not difficult to find right-angled triangles that work for $N = 2$ and $N = 4$, but not all numbers are so easy. Indeed, there is no triangle that works for $N = 6$.

However, there is no reason to stick to triangles. The same argument would work for any figure that can be dissected into smaller figures (all of the same size) of exactly the same shape. Such figures were first studied by mathematician Solomon Golomb, who gave them the charmingly apt name of rep-tiles. They were then popularised by Martin Gardner in one of his excellent columns.

Some rep-tiles are not too difficult to find, but others are clever and beautiful:

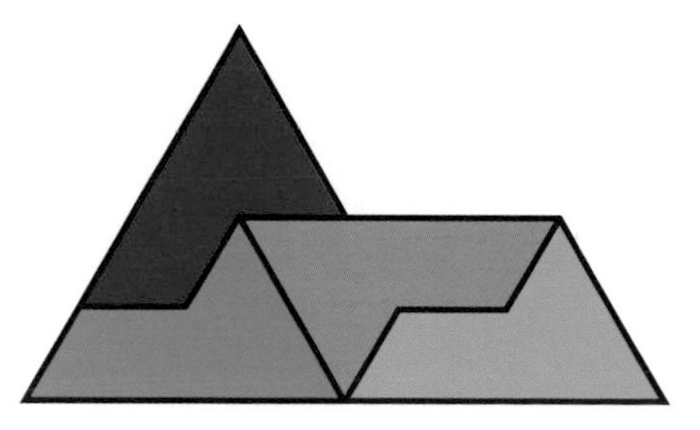

Of course, if we're willing to go down a dimension, then the simplest rep-tile is a plain old line segment.

Summing the lengths of the sub-segments, we obtain our first geometric sum (for $N = 5$), exactly as we did with Federation Square's triangles. Still, Federation Square does it with much more style.

Puzzles to ponder

A rep-tile that splits into N identical pieces is called rep-N. So, the tiles discussed above are rep-5, rep-3, and rep-4.

1. Find a triangle that is rep-2.
2. Find a triangle that is rep-4.
3. Given any positive whole number $N \geq 2$, find a rectangle that is rep-N.

CHAPTER 20

Excavating a mathematical museum piece

Melburnians are very familiar with a curious corner, sticking out of the Swanston Street footpath, where the Melbourne Museum used to reside. Its official name is *Architectural Fragment*, and it was created in 1992 by Victorian sculptor Petrus Spronk. It is striking, and it appears vaguely mathematical—we'll excavate the mathematics.

We can think of the *Fragment* as the corner of a cube. What you see when you walk around are three right-angled triangles making up the corner. This brings to mind *Pythagoras's theorem*: we are all familiar with the mantra "A squared plus B squared equals C squared", relating the two shorter sides of a right-angled triangle to the longer side.

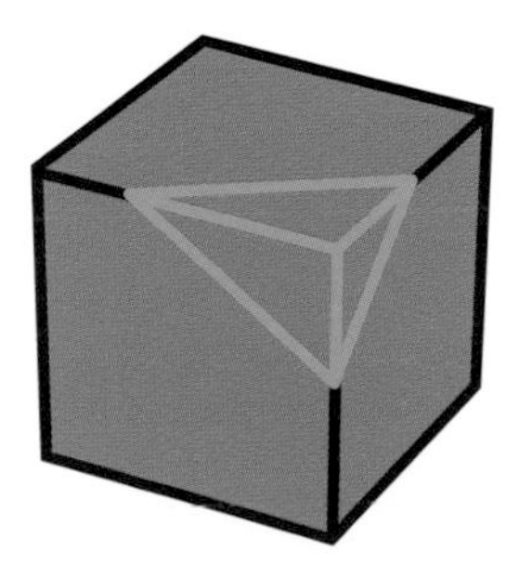

Actually, "Pythagoras's theorem" may be a misnomer. Pythagoras lived around 500 BC, but his theorem was known to the Babylonians, over 1000 years earlier. True, the Babylonians seemingly didn't know how to prove Pythagoras's theorem. But, there is also no strong evidence that Pythagoras knew how!

What the Babylonians did know about were *Pythagorean triples*: right-triangles with whole number sides. The simplest such triple is 3-4-5, but there are many others. Such triples are not just mathematically pleasing, they are also of practical use. For example, if you are landscaping and need to make a right angle, then take 12 meters of rope and form a 3-4-5 triangle: the large angle will be exactly a right angle.

And this exact technique was applied to make the *Fragment.* The architect Spronk took a 3-4-5 rope triangle and fixed the 5-meter side to the footpath. He then had a friend hold up the opposite corner and move it up and down until it "looked right". The sculpture was then constructed from bluestone around a steel frame, and cemented into the footpath.

There is another, much less familiar, Pythagorean gem hiding in the *Fragment.* Think of the corner as a pyramid, made up of three right-angled triangles and a triangular base (which will not be right-angled). Then the areas A, B, C, and D of these triangles are related by the simple formula

$$A^2 + B^2 + C^2 = D^2.$$

This beautiful result is known as *de Gua's theorem*, and it's about 300 years old: a baby compared to Pythagoras's theorem.

But is de Gua's theorem really Pythagorean? One way to think of a right-angled triangle is as the corner snipped off a square. Then de Gua's theorem is simply the Pythagorean theorem but one dimension higher, for the corner snipped off of a 3-dimensional cube. And you can go further: just put on your 4-dimensional goggles, snip the corner off of a conveniently located 4-dimensional hypercube, and you'll see what we mean.

Puzzle to ponder

Can you think of four positive integers A, B, C, and D such that

$$A^2 + B^2 + C^2 = D^2.$$

CHAPTER 21

AcCosted at Healesville

One of Victoria's fabulous family offerings is Healesville Sanctuary. It's a great place to view native Australian wildlife, and one of your Maths Masters and his very willing family just made the trip for the umpteenth time. However, on this occasion there was also a wonderful surprise: amidst the exotic Australian wildlife was hiding a rare and beautiful mathematical creature.

Located at Healesville Sanctuary is the Australian Wildlife Health Centre. It's a very impressive training and research hospital, as well as being a fascinating educational component of the Healesville tour. Visitors are able to talk to the vets and nurses, and to watch operations being performed on the animal patients.

On his recent trip, your Maths Master was witness to some involved wombat dentistry, and his squeamishness encouraged him to search around for more pleasing sights. That was very fortunate, because he happened to glance up at the roof of the Health Centre: it was curved in a beautiful and elaborate manner, and it seemed strangely familiar. And then he figured it out.

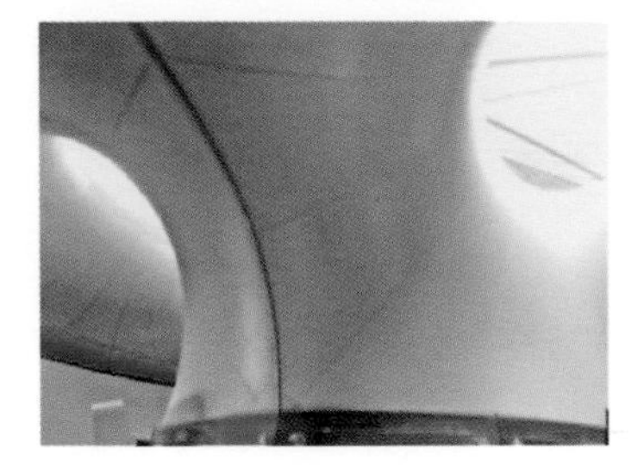

The Health Centre roof is a close mathematical relative of *Costa's surface*, discovered by Brazilian mathematician Celso José da Costa in 1982. Costa's surface created a huge buzz, and was named *Time Magazine*'s Surface of the Year in 1985. (Well, it would have been if *Time* had a mathematics section.) We'll try to explain the reason for all the excitement.

A computer rendering of the main part of Costa's surface appears below. However, unlike the computer graphic (and the roof of the Health Centre), Costa's surface continues beyond the orange boundary curves, forming an infinite surface without edges—in the lingo, Costa's surface is said to be complete.

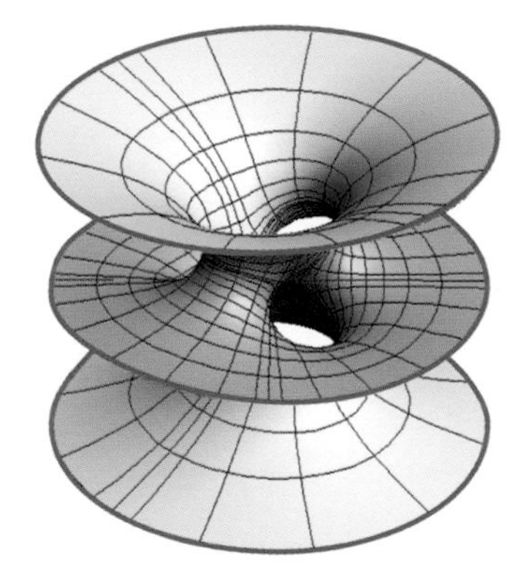

So, what's so special about Costa's surface? To begin, it is what is known as a minimal surface, which is the mathematical model of a soap film. Imagine a small loop embedded into Costa's surface, as pictured below.

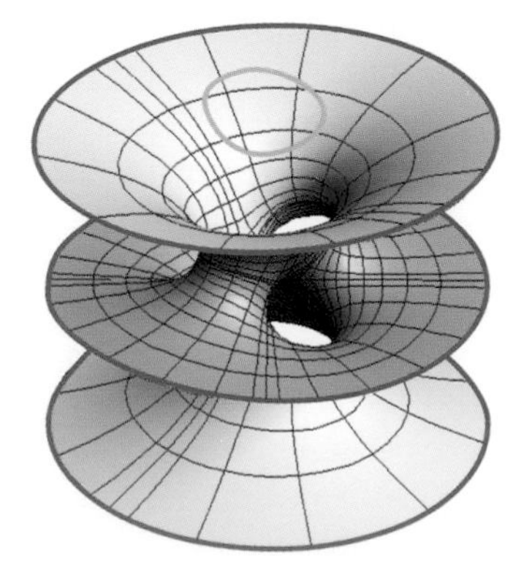

Now, think of the loop being made of wire, and imagine dipping the wire in a soap solution. Then the soap film that would form is exactly the piece of Costa's surface within the loop. Mathematically, this is saying that little pieces of Costa's surface have least area amongst all imaginable surfaces with the same edge.

Mathematicians had been discovering and studying minimal surfaces for centuries, but until recently they knew of very few complete minimal surfaces. In fact, amongst complete minimal surfaces with the extra property of finite topology (meaning roughly that the surface has a finite number of holes), only three types were known: a boring flat plane; the staircase-shaped helicoid; and the tube-shaped catenoid.

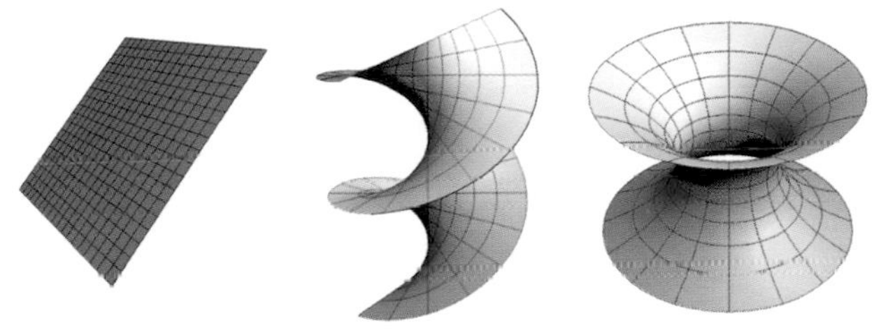

And then came Costa's surface. But if Costa discovered his surface in 1982, why did it take until 1985 for all the cheering to begin?

Costa used a technical method (involving complex numbers and integral calculus) to write down the equations for his surface. Consequently, Costa's surface was very difficult to analyze and to picture; the tricky question was whether Costa's surface would curve around and run into itself. Such self intersecting minimal surfaces are quite common and are much less interesting; a surface which avoids these self-intersections is referred to as embedded.

In 1985, mathematicians David Hoffman and Bill Meeks employed (very early) computer graphics to try to understand Costa's surface. Their graphics were incredibly grainy, but Hoffman and Meeks saw enough to be convinced that Costa's surface was indeed embedded. The graphics also indicated that Costa's surface had a number of symmetries, which were not obvious from Costa's equations.

Hoffman and Meeks then returned to take a careful look at Costa's equations. With the symmetries to guide them, they were able to prove conclusively that Costa's surface was embedded—that it was indeed a new complete minimal surface.

The discovery and analysis of Costa's surface opened the floodgates. There has since been discovered an incredible zoo of minimal surfaces. The Health Centre roof is modelled on what is known as the *Meeks–Hoffman–Costa surface*, and features an order 3 rotational symmetry; it is closely related to Costa's original surface, and it was proved to be embedded soon after.

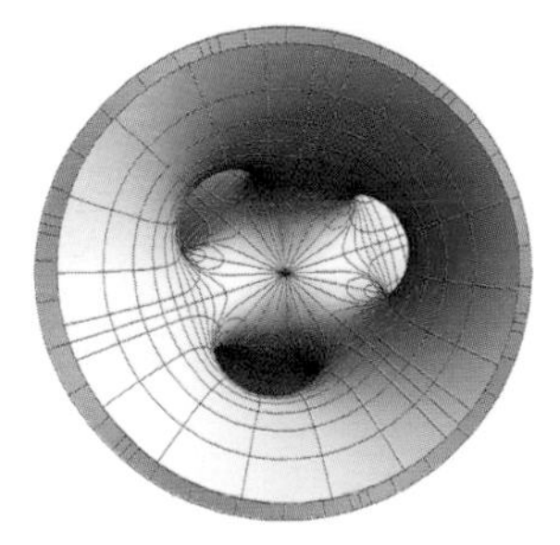

Not surprisingly, turning Costa's beautiful surface into the Health Centre's stunning roof was anything but straightforward.

Of course, nowadays computer graphics are much easier and cheaper, and beautiful graphics of minimal surfaces (and everything else) are very easy to find. In fact, the images for this chapter were created with the excellent free software 3D-XplorMath.[1] We urge you to download the program and explore for yourself!

So, there you have it. Healesville's Health Centre is yet another beautiful addition to our growing and growing and growing collection of Melbourne's mathematical architecture. What an amazing city!

Puzzle to ponder

The catenoid can be thought of as a deformed sphere from which two circles have been cut out. Costa's surface can also be thought of as a deformed version of a simple surface, from which three circles have been removed. What is that surface?

[1] `3d-xplormath.org`

CHAPTER 22

The Klein Bottle Beach House

Two leisurely weeks of school holidays, and what better place to relax than on the Mornington Peninsula? There's hot springs and wineries and beautiful beaches. And, there's a very mathematical beach house.

Rye's stunning Klein Bottle House was constructed in 2008. This award winning beach house is the creation of Melbourne architects McBride Charles Ryan. (The lovely photographs are by architectural photographer John Gollings.)

The house is absolutely amazing, but where's the math? Where's the bottle? Who is Klein? We have some explaining to do.

The *Klein bottle* is named after its discoverer, the 19th-century German mathematician Felix Klein. It is a famous mathematical surface, a close relative of the mega-famous Möbius strip. Turn the page for a photo of a glass Klein bottle (available from the very quirky ACME Klein Bottle Company[1]).

Klein bottles needn't be made of glass. The discerning football fan can acquire snug and stylish Klein bottle beanies. (The example below, with matching Möbius scarf, was created by our very own slave-knitter, also known as Grammy Maths Master.)

Now for some details. To make a Klein bottle, begin with a rectangle and glue two opposite edges together to make a cylinder. Bend the cylinder around so it

[1] www.kleinbottle.com

passes through itself, and finally glue the two circular ends together. (The pictures below were created with the 3D visualisation software *JavaView*.)

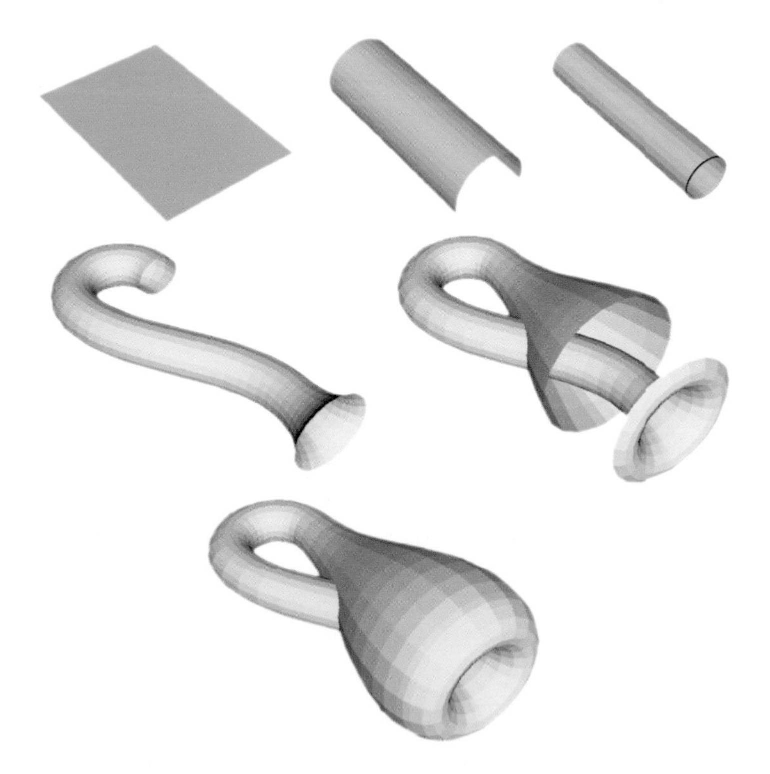

There is an obvious objection to the above procedure: since when are we allowed to have a surface simply cut through itself? Mathematicians are not at all bothered by this, blithely referring to self-intersecting surfaces as *immersed*. However, we admit that such surfaces are somewhat dissatisfying. It also turns out that for the Klein bottle these self-intersections are unavoidable. (Well, they're avoidable if you have a fourth dimension handy: a case where the cure is probably more puzzling than the disease.)

A second method of creating the Klein bottle demonstrates that it has a Möbius strip hidden within it. Recall that a Möbius strip is constructed by taking a long rectangular strip, giving it a half-twist and joining the ends together. The resulting surface has a single edge. Now, if we widen the Möbius strip in just the right manner, the edge will close in on itself and the end result is our Klein bottle.

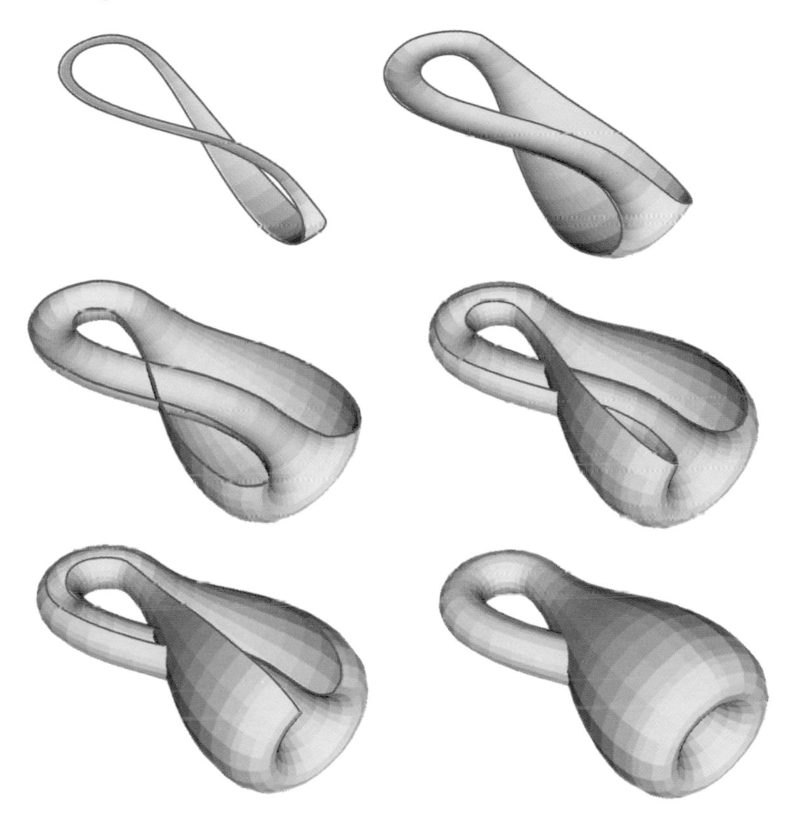

It follows that the Klein bottle is one-sided just as is the Möbius strip. A bug crawling on the "inside" of the Klein bottle will wind up on the "outside".

What about the Klein Bottle House? On the face of it, a one-sided self-intersecting surface is an unlikely model for a beach house. If nothing else, the heating costs would be astronomical.

Fortunately, the Klein Bottle House is not really a Klein bottle. Nonetheless, having viewed the many pictures of this stunning house, we can attest to its very Klein bottle-ish feel.

The architects Rob McBride and Debbie-Lyn Ryan were apparently inspired by origami Klein bottles.

So, our hopes of living in a real Klein bottle house have been dashed. But your Maths Masters have no quibbles with McBride's and Ryan's very stylish interpretation. We'd both be delighted to move into one, preferably somewhere on the beautiful Mornington Peninsula.

Puzzle to ponder

A Klein bottle can also be made by gluing two Möbius strips together along their edges? Can you see two such Möbius strips inside a glass Klein bottle?

CHAPTER 23

Waiting for the apple to fall

The next time you are looking for some mathematical inspiration, head out to Monash University in Melbourne's east. There, you will find the Newton apple tree.

That Newton? That tree? Yes, Monash indeed has a tree linked to the famous story of Sir Isaac Newton and his discovery of gravitation. But no, it is not a case of some fanatical Aussie mathematicians abducting Newton's tree.

First, we must qualify the myth: it is definitely not the case that Newton hit upon the concept of gravitation after being whacked on the head by a falling apple. It is also misleading and disrespectful to reduce Sir Isaac's years of intellectual struggle to one moment of genius.

But there is also truth to the story. Newton did relate that it was while pondering in a garden that he made a critical breakthrough: he realised that the force that makes objects fall might also be acting upon the Moon, and that this may explain the Moon's orbit. And, in more than one telling of the story, it is a falling apple that inspired Newton's thoughts.

So, there probably was a Newton apple tree, and there are actually a number of trees competing for the honor. The strongest claim by far is for the apple tree that grew at Woolsthorpe Manor, Newton's birthplace. This grand old tree toppled over in a storm in 1816, with part of it then being made into a chair. However, the tree apparently rerooted and in this reincarnation still happily survives.

After Newton's death, the owners of Woolsthorpe Manor transferred cuttings of the tree to Belton Park, Lincolnshire, a few miles away. From there, cuttings were later transferred to the National Fruit Research Station in East Malling, Kent.

There are now many supposed Newton trees, all claiming to be direct descendants of the original. However, recent DNA fingerprinting has revealed many of these trees to be impostors. We are happy to report that Monash's tree is indeed a direct descendant of the likely original Newton tree, via the East Malling stock.

The Monash tree was planted in 1972, unlabeled and situated in a secluded corner of the Engineering courtyard. Apparently there had been fears of vandalism, although it's not clear from whom: perhaps some rioting fans of physicist Robert Hooke, Newton's fierce rival? In any case, the tree was then moved to its current location in 1975. Below is a photo, showing curator John Cranwell and his staff lovingly carrying out the transplantation.

Monash University Archives, IN705.

Have you ever seen an apple drop from a tree? Neither have we. But we did surreptitiously sample one of the apples from the Monash tree. The tree is a variety known as Flower of Kent, and our sampling suggested why Newton may have left the apples undisturbed, to drop on their own: Flower of Kent apples taste terrible! They may be great apples for inspiring genius, but they are pretty much inedible.[1]

Puzzle to ponder

Sir Isaac is making an urgent delivery of apples: three huge, inspiring apples. He comes to a rickety bridge. The bridge can just carry Newton and two of the apples, but the weight of the third apple would be way too much. In a flash of brilliance, Sir Isaac decides to juggle the three apples while crossing the bridge. He reasons that one of the juggled apples will always be in the air, and that the bridge can take the weight of the other two apples. Is Sir Isaac being a genius, or will he and the apples and the bridge all collapse into the water below?

[1]We're not being all that fair, since the Flower of Kent is considered to be a cooking apple.

Part 5

A Lotto gambling

Australians love gambling. It is said that Australians will bet on two flies crawling up a wall, and we believe it. Any country that can make gambling on the flip of a coin into a cultural icon (Chapter 24), well they can find a way to wager on pretty much anything.

This would all be charming except of course that it isn't. The national and state governments have been all too happy to permit hucksters to fleece the sheep, as long as the governments receive their share of the fleece. Many, many lives have been ruined. And of course none of the mathematics that follows is of any use whatsoever. Compulsive gamblers are too busy imagining that God is on their side to be bothered with calculating or respecting probabilities and expectations. Such is life, larceny, and stupidity.

CHAPTER 24

A bucketful of two-up

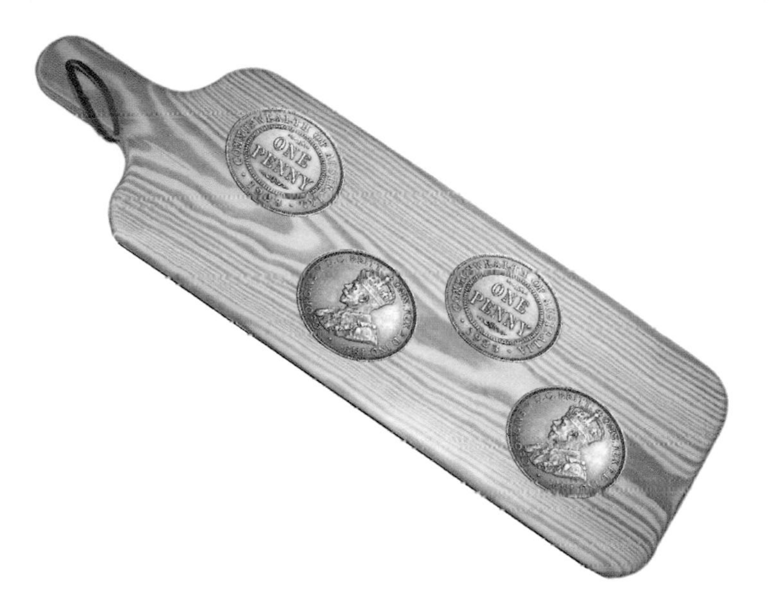

Anzac Day[1] of course brings with it the traditional game of two-up. We have been pondering this, and we got sidetracked. The result is our new game: bucket-up.

For the diggers,[2] with limited resources and much more pressing concerns, two-up had to be simple: two coins were thrown, and bets were placed. Standard bets were "Heads" or "Tails", where both heads or both tails came up, respectively.

The probability of a head for each coin was $1/2$, so the probabilities of Heads was $1/2 \times 1/2 = 1/4$, and similarly for Tails. The remaining possibility, of one head and one tail, was known as "Odds" and had a $1/2$ chance of occurring. Traditionally, Odds was not bet upon, and it simply meant "toss again".

But if Odds didn't count, this left Heads and Tails as the equally likely outcomes. So, why not just flip a single coin? Possibly because using two coins made it harder to cheat. Or possibly two coins made the game more entertaining, by adding a delaying tension when Odds occurred.

Whatever the reason, if two-up is better than a single coin, then four-up must be better still. But, with four coins or more, you have to be careful.

[1] Anzac Day, celebrated on April 25 each year, is essentially Australia's version of America's Memorial Day, to remember those who served and died in wars.

[2] Australian slang for "soldiers".

There are many ways four coins can come up. What about Odds (where you bet that exactly two heads and two tails will occur)? Would you take that as a fair 50-50 bet?

Don't fall for this. In fact, it is a well-known con. There are 16 ways in which the coins can come up. And of these 16 ways, only six give equal heads and tails. So the probability of Odds is $6/16 = 3/8$.

The chances of Heads or Tails in four-up are much smaller, only $1/16$ each. The more coins we use, the smaller are the chances that Heads and Tails and Odds occur. But the way these chances diminish is paradoxical.

For example, how many coins do you think are needed so that Odds is less likely than winning the big prize in Tattslotto?[3] We'll give you two clues. First, 30 coins is enough to ensure that Heads is less likely than winning Tattslotto. Second, count the number of grains of sand that fit in an olympic-sized swimming pool.

We now give you the logical extreme of two-up: the game of bucket-up. What you do is take a bucket of coins and throw them up in the air. It turns out that the chances that an even number of heads turn up, or that an odd number of heads turn up, are both exactly $1/2$.

The fact that bucket-up is truly a 50-50 game is not obvious, and we'll leave you to puzzle over that. More importantly, is bucket-up really an improvement over two-up? For the definitive answer, gather some children, then ask them whether they would prefer to play two-up or bucket-up.

Puzzles to ponder

1. Prove that bucket-up (where you can guess the number of Heads to be even or odd) is a 50-50 game.
2. The chances of winning Tattslotto are about 1 in 8 million. How many coins are needed so that Odds (the same number of heads and tails) is less likely than this? Of course, the question only makes sense if we consider beginning with an even number of coins.

[3]The most well-known lottery in Australia.

CHAPTER 25

We get a Lotto calls

A while ago there was a big jackpot ready to go off in Oz Lotto.[1] As is standard when these jackpots appear, mathematicians received a succession of calls from the media.

We would have thought that a simple "Buckley's chance"[2] aptly sums up the likelihood of any given person winning. However, it seems that the public loves to hear official mathematical versions from official mathematicians. So, as we have in the past, your Maths Masters obliged.

First, let's get the Buckley part out of the way and figure out the chances of our lottery ticket winning the jackpot.

On our Oz Lotto ticket we mark off seven numbers from 1 to 45. Then, 45 numbered balls are placed into a machine and seven balls are drawn out. Our ticket will win the jackpot (Divison 1) if we have correctly chosen the seven drawn numbers.

To figure out the chances of doing this, we can calculate ball by ball. For the first ball there are 45 numbers that might come up, and seven of those would match a number on our ticket. That means we have a 7/45 chance of our ticket still being in the running after the first ball is drawn.

Suppose we have correctly chosen the first ball. There are then 44 balls left and we have six numbers left on our ticket to match them. So, at that stage we have a 6/44 chance of our ticket surviving the second ball.

Working through all seven balls, we can see that the chances of choosing all seven numbers correctly—and so being able to retire to a tropical island—is

$$\frac{7}{45} \times \frac{6}{44} \times \frac{5}{43} \times \frac{4}{42} \times \frac{3}{41} \times \frac{2}{40} \times \frac{1}{39}.$$

That works out to about 1 chance in 45 million. (Actually, 1 chance in 45,379,620, for those who want to know the precise value of a Buckley.)

Newspaper reports of lottery jackpots always include comparisons to unlikely events, such as being hit by lightning. Here is one stunning comparison that was suggested to us by a colleague.

[1] One of Australia's bigger lotteries.

[2] A popular Australian expression meaning little or no hope.

Below is a diagram of the MCG,[3] and at left full forward you can see a black dot. Now, imagine dropping a dart from high above the MCG and assume it hits somewhere randomly on the field. What are the chances of the dart hitting that black dot? It's clearly not very likely, and the chances of winning Oz Lotto are much less.

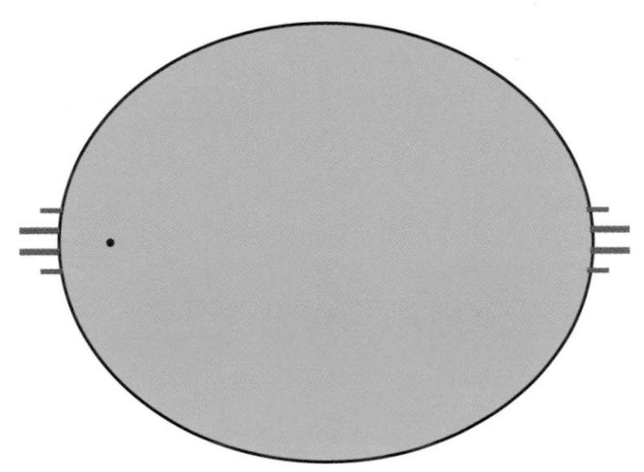

The dot we have drawn represents a circle of about 2 meters in diameter. However, the chances of winning Oz Lotto correspond to hitting a circle that is 2 *centimeters* in diameter, about the size of an Australian 10 cent coin. To put it another way, your chances of winning the jackpot in Oz Lotto are about 1/10,000 the chance of hitting the "large" dot we've actually drawn. (The black dot that would actually represent winning Oz Lotto is too small to show up in print.)

OK, so after all that, it is clear that our chances of winning Oz Lotto are about what we originally suspected: Buckley's. However, there's another interesting question we can ask.

As it happens, there was no Division 1 winner in that big Oz Lotto draw. Was that surprising? What were the chances that *someone* would win?

It seems that the actual number of lottery tickets was around 50 million. But let's cheat a bit and assume that there were 45 million tickets; that is, we assume there were just enough tickets to cover every combination. Of course, many combinations would have been repeated, and so many other combinations would not have been covered at all.

Now we consider each ticket in turn. The chances of the first ticket *not* winning were (45 million − 1)/(45 million). And, the same was true for all 45 million tickets. To find the chances that *none* of the tickets would win, we multiply the individual chances, and that gives

$$\left(\frac{45 \text{ million} - 1}{45 \text{ million}}\right)^{45 \text{ million}}.$$

We can easily throw that into a calculator, but there is a very interesting alternative. Rearranging the fraction slightly, we can rewrite the expression as

$$\left(1 - \frac{1}{\text{huge}}\right)^{\text{huge}},$$

where "huge" happens to equal 45 million. And now, this may ring a bell.

[3]The Melbourne Cricket Ground, a large and much-loved ground for playing cricket and Aussie rules football.

The end result of the quantity $(1+1/\text{huge})^{\text{huge}}$, as "huge" gets into the zillions, is defined to be the famous number e, which is about 2.7.[4]

It takes some fiddling (which we'll leave for you as a puzzle) but it turns out that our new expression, $(1-1/\text{huge})^{\text{huge}}$, is very close to the reciprocal $1/e$, which is about 0.37. That's the chance that no one would win the jackpot, leaving about a 63% chance of a winner.

So with that many tickets entered, we'd expect that there'd be a winner, but it's not a huge shock that there wasn't. And of course, with everyone and their dog buying tickets for the next Oz Lotto draw, it was no surprise at all that the jackpot then went off.

Puzzles to ponder

1. Supposing that 90 million tickets are entered into Oz Lotto, what are the chances that someone will win?
2. Prove that $(1-1/\text{huge})^{\text{huge}}$ is approximately $1/e$.

[4]Just for the record, e is *not* "Euler's number", despite what pretty much every senior mathematics textbook claims.

CHAPTER 26

Winning the Lotto by the Melbourne method

8 + 11 + 13 + 15 + 19 + 20

=

$ 27,000,000

On February 15, 1992, a syndicate held the single winning ticket for the Jackpot of the Virginia Lottery, worth US $27 million. The syndicate was headed by Romanian-born Melburnian, Stefan Mandel, who had previously won 13 lottery jackpots. How incredibly lucky!

Of course, the explanation has nothing to do with luck. Rather, Mandel and his associates would simply buy tickets covering all possible combinations, thereby guaranteeing they held a winning ticket. Obviously, this strategy can only make sense if the jackpot is large, so that the combined prizes will hopefully exceed the cost of all those tickets. However, even when the jackpot is huge, the strategy is still a gamble, since there is always the chance of having to share the jackpot with winners outside the syndicate. So, what is called for is a really, really huge jackpot, with as few participants as possible. The Virginia Jackpot in 1992 was the state's largest up to that point, making it a perfect target for Mandel.

Despite Mandel's expertise, things almost went wrong. The Virginia lottery required the picking of six numbers out of 44, meaning the syndicate had to fill out and purchase a total of 7,059,052 individual tickets. The 35 people assigned this enormous task managed to buy around five million tickets, but this still left much more of a gamble than Mandel had planned upon. On this occasion, fortune favored the brave.

What made the Virginia lottery a logistical nightmare was the fact that each combination had to be covered with an individual entry. By contrast, many lotteries offer system tickets, covering numerous combinations with the one entry. For example, Australia's Tattslotto requires you to pick six numbers out of 45, and a System 8 ticket allows you to choose eight numbers: this one ticket costs $19.90 and gives you all 28 six-number combinations from these eight numbers. Or, for those who prefer a slightly larger flutter, there is the System 20 ticket, covering 38,760 combinations and costing a mere $27,537.05.

When available, system tickets streamlined Mandel's operation, but even then there remained some very tricky mathematics. For example, if we were to apply Mandel's strategy to Tattslotto, then we would buy System 20 tickets to cover all six-number combinations. But how many tickets would we need? No one knows!

There are 8,145,060 combinations in Tattslotto, and so the arithmetic says we need at least 211 System 20 tickets. But in fact we will need more: in trying to cover all combinations, our System 20 tickets will unavoidably overlap.

However, Mandel didn't require the very best mathematical solution, as long as his method didn't involve too much overlap. Indeed, he succeeded so well that many lotteries responded with rules to counter such schemes. Not that syndicates in themselves lower the profits for the lottery owners, but apparently the regular suckers—sorry, participants—were becoming annoyed.

You can also apply Mandel's idea in trying for smaller prizes. For example, suppose you want to choose enough tickets in Tattslotto to guarantee having three of the six numbers: that gives you a good shot at a smaller prize. Again, the mathematics is difficult, and no one knows the minimum number of tickets required to achieve this. However, it is known how to choose 154 tickets that suffice, for a cost of about $110.

But what's the point of such small-prize hunting? It doesn't make much sense outlaying $110 in an attempt to win a prize of about $50; in the long run, you can still only expect to get back about 60% of what you gamble, just like everyone else. And, these 154 choices are no magic ticket to winning larger prizes. It's all pretty silly. Nonetheless, there are hucksters—sorry, businessmen—selling such schemes, and there are suckers—sorry, participants—buying them.

And finally, what became of Mandel? His story ends sadly. The Virginian authorities, who were not at all pleased with Mandel's coup, launched an inquiry involving all manner of American and Australian law enforcement. Although no wrongdoing could be proven (since obviously there was none), the U.S. tax department imposed a 30% tax upon the syndicate, claiming their winnings amounted to professional income. Eventually the members of the syndicate fell out, and in 1995 our hero Mandel was declared bankrupt.

Puzzle to ponder

If there was such a thing as a System 44 ticket for Tattslotto, what's the minimum number of these super-tickets required to cover all possible six-number combinations?

CHAPTER 27

The super-rigging of poker machines

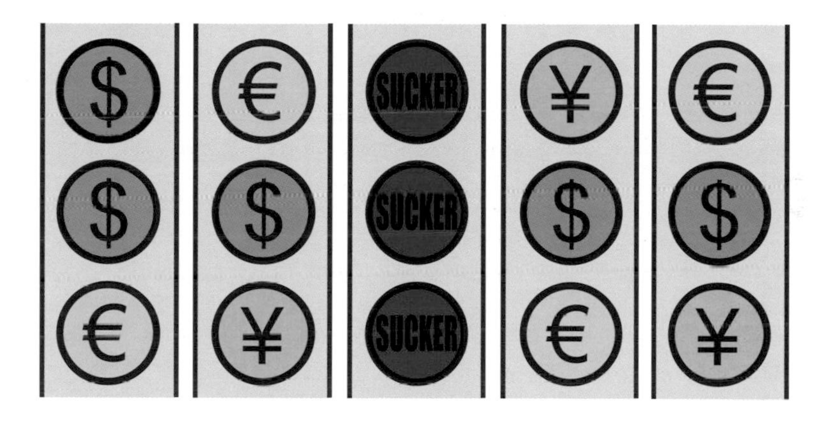

Are poker machines rigged?[1] Of course! Even the most optimistic gambler knows that the odds are stacked against them. But it is worse than that: poker machines are super-rigged.

Gambling is gambling. For those who engage, its unpredictability is part of the charm and the excitement. But for any gambling game, mathematics can still guide your expectations.

If a game is fair, then winning or losing comes down to luck, and in the long run you expect to get back close to what you gambled. Of course, no one expects pokies to be fair. In fact, of each $100 gambled, you can only expect to get back about $90.[2] It is in this sense that the pokies are rigged. This rigging is also more costly than it may appear, because it is likely that the same money will be gambled over and over, with a fraction lost each time.

But how can poker machines be super-rigged? The trick is in the psychology—to make it appear that the chances of winning are greater than they really are. Such scams have a proud tradition in carnival games. The comparable super-rigging of some poker machines has been documented by Melbourne barrister Tim Falkiner.

To illustrate super-rigging, consider the following simple game. Throw a pair of dice. If the sum is 10 or higher, then you win the Big Jackpot, with smaller prizes for other totals. This game can easily be rigged to give less than 100% money returned, simply by adjusting the sizes of the prizes.

Now consider the same game, but super-rigged with dodgy dice. Imagine we swap the 4 and the 6 on one die for the 3 and the 5 on the other: so, the sides of the dice are now 1-2-3-3-5-5 and 1-2-4-4-6-6.

[1] Australians refer to slot machines as poker machines, or pokies.

[2] In Australia a minimum long-run return of around 90% is guaranteed by law.

The dodgy dice together have the same numbers as normal dice. This implies that the average sum of a roll of the dice has not changed, giving the psychological impression that prize-winning totals are just as likely.

But actually, the chance of winning the Big Jackpot has decreased dramatically. With normal dice there are six of the 36 possible rolls that will sum to 10 or higher; the rolls are indicated in the left diagram below.

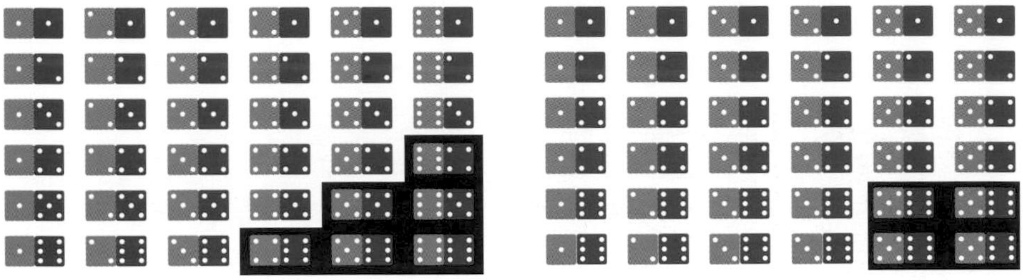

With the dodgy dice, there are now only four such rolls, as indicated on the right. So, the chances of winning the Big Jackpot have gone down from 6/36 to 4/36—a 33% reduction. And, cunningly, the chance of "just missing" by rolling a 9 has doubled, from 4/36 to 8/36.

It is exactly this type of super-rigging that is programmed into the pokies. Think of the five poker machine wheels as 30-sided dice. Some wheels are starved of Jackpot symbols, which is then disguised by loading a few more Jackpot symbols on other reels. The consequence is that winning the Big Jackpot is much less likely than it appears. And, the chances of "just missing"—encouraging another go—is much more likely.

The use of super-rigged poker machines is incredibly sneaky. Is it also illegal? Tim Falkiner has argued that if such machines are "deceptive", then they could well be banned under the Trade Practises Act.[3] However, the Australian Competition and Consumer Commission is apparently unconvinced that such machines are deceptive. We feel compelled to ask: what would be?[4]

Puzzle to ponder

It is quite well known that on any given roll of two dice, a total of 7 is most likely. Now imagine you are playing the following game: A pair of dice is thrown over and over. You want a total of 7 to appear twice. Your opponent is waiting until a total of 6 and a total of 8 have appeared. Would you bet that you can get there first?

[3]An Australian consumer protection law, which in many instances has turned out to be very powerful. Of course, America has no comparable history of consumers being fleeced by dodgy companies, and so Americans have no need for such a law.

[4]In late 2016 a pokies player brought a lawsuit against a casino and poker machine manufacturer, arguing this very point. At the time of writing, the case has not been settled.

CHAPTER 28

How much is a $100 free bet worth?

It's almost football time again. And it's time to make a little wager on our beloved Saints winning the premiership.[1] It turns out that a number of sports-betting websites offer a free bet as inducement to open an account with them. However, the value of their offer is surprisingly tricky to determine.

Suppose you are offered a $100 free bet. This is definitely not the same as being given $100: you have to bet on something. But there are other differences as well.

For example, suppose you want to bet that the Saints will win the toss of the coin in their first match. With a normal bet, you would hand over $100. Then, ideally, a successful bet would return $200, the $100 you wagered plus the $100 you won. In reality, you'll receive slightly less than fair odds. So you might win about $95. But, let's ignore that for now.

With the free bet, you don't give over the $100 stake, and then you usually don't get it back either. So, a winning free bet would mean you receive only the winnings of $100.

Imagine now that you have two free bets with two different companies. You can then use one to bet that the Saints will win the toss and the other to bet that they will lose. Then, whatever happens, the two bets result in your receiving $100.

So, it seems each $100 free bet averages to a worth of about $50, minus a little for the actual odds being unfair. What is amazing is that you can actually do better. This was pointed out to us by our sports journalist friend, Gary Watt. Gary, who has no degree in mathematics, calmly explained it all in the face of our smug skepticism.

Imagine that the 18 teams in the AFL have an equal chance of winning the premiership. Then a $100 free bet on the winning team would win about $1700.

[1] St Kilda, an Australian rules football team, which has a long and proud history of losing.

Now imagine you have eighteen free bets, and you use one on each team. Then, whoever wins, you'll get back about $1700. So each free bet averages to a worth of $1700/18—about $94!

The remarkable conclusion is that on average a free bet is worth more if it is less likely to win. This makes it perfect for a flutter on our beloved and beleaguered Saints.

Reference: Marty Ross, *How much is a $5 betting coupon worth?*, Math Horizons, September 2008, 30 (available on our website `www.qedcat.com/articles`).

Puzzle to ponder

When it first opened, the Crown Casino in Melbourne charged for admission. In return, the patron was given a $5 betting coupon: The $5 coupon could be used to make any standard casino bet. If the bet succeeds, the patron receives their normal winnings; Win or lose, the Casino collects the coupon. What was the value of such a coupon? (It was pondering this exact question that inspired our column.)

Part 6

Keeping the bastards honest

Australian politics is dominated by two political groups, the nominally liberal Labor party and the phenomenally conservative Liberal-National coalition. Actually, they're not that conservative, at least not by American standards; Australian conservatives only rarely lie the country into meaningless wars, and they don't often advocate torture.[1] Nonetheless, Australia has no shortage of nasty, neoliberal cranks.

It is well known that any voting system can be "proven" to be unfair (see Chapter 30), and in practise the shortcomings of different countries' systems are easily determined. Of course they're not typically as bad as America's, which is so devoid of honesty, fairness, and plain common sense, that it can described as "democratic" only for the purposes of black humor.

The Australian voting system is also demonstrably unfair, in silly and easily correctable ways. However Australia's system is not so unfair that smaller political parties don't have some power. Such parties love to talk about keeping the bastards honest. They do so only very weakly, but at least they help to shine a light on the dishonesty. And, they provide us with plenty to write about.

---

[1] This was written before the hilarious Donald Trump was elected president. We thought to revise this introduction, but the words simply failed us.

CHAPTER 29

Tomfoolery and gerrymandering

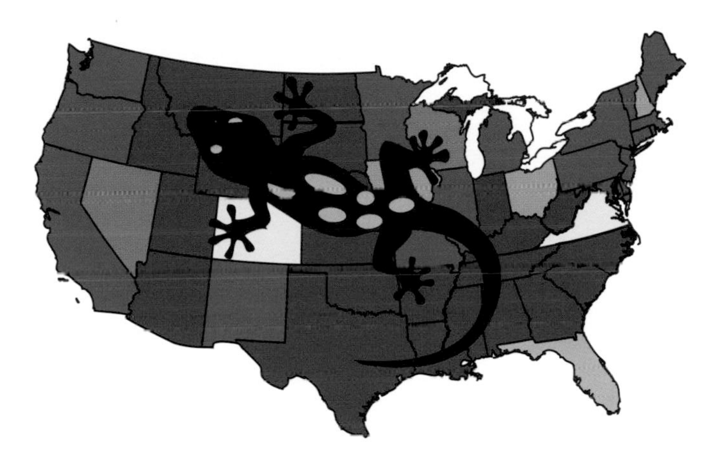

You're up for election and you're up against it. Your financial policies are a mathematical fantasy, and your social policies are from another century, if not another planet. How can you possibly win?

Luckily, in America at least, you still have plenty of options. For example, you could simply lie about your policies, and you might have a "news" network to actively assist you. Then maybe a like-minded, ethically deficient court could assist you in buying the election. And if that's not enough, you could simply deny millions of undesirable voters the right to vote against you. America truly is the land of opportunity (for hucksters).

The above are all tried-and-true methods. However, if you're a genuine American traditionalist, then you'll seek to use one technique above all: gerrymandering.

Gerrymandering is the practise of redrawing the boundaries of electoral districts to assist the re-election of a political party. To illustrate, consider the following election being fought between the red party and the blue party.

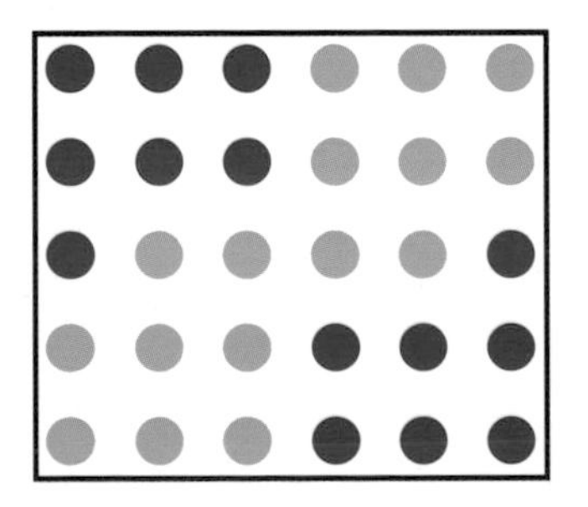

The country up for grabs consists of 14 red voters and 16 blue voters, and so the blue candidate should win. Unfortunately, not many elections are decided in such a clear and reasonable manner.

Often, as is done in Australia, the state or country is divided into a number of electoral districts. Each district is decided by majority vote, and the party that wins the most districts wins the election.

This electoral system for determining a country's leader is patently ridiculous. It's a good reason why Australians shouldn't be overly smug when considering other countries' electoral failings. But more than ridiculous, the electorate system is notoriously open to manipulation.

Returning to the battle between the blue and red parties, let's imagine their country is divided into three electorates, each containing ten voters, as pictured.

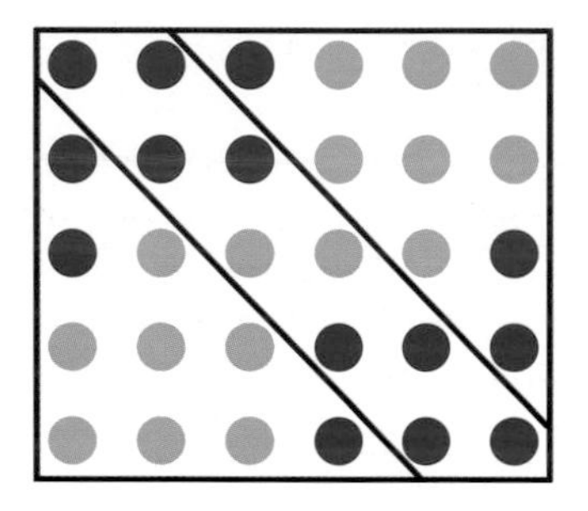

The blue team will win in two districts and the red team in just one, which is still bad news for the red team. However, if the red team is in power, then they could redraw the district boundaries to suit themselves:

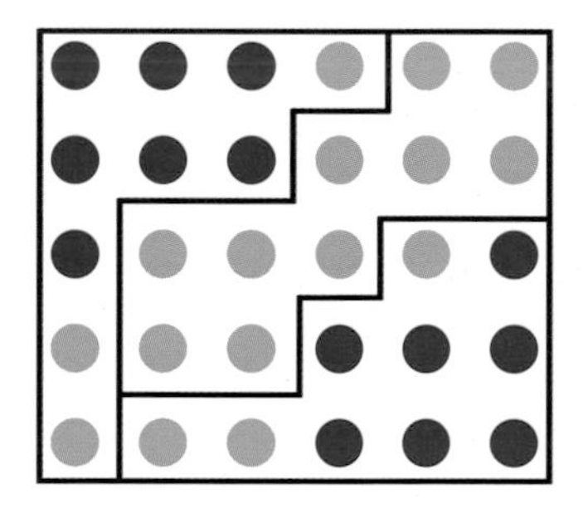

Each district still contains ten voters. However, by packing most of the blue voters into one district, the cunning red team can ensure they win two districts, and so the election.

Can such blatantly antidemocratic scheming possibly go on? Amazingly, yes. Indeed, gerrymandering seems to be as American as apple pie.

The practise is named after Elbridge Gerry, governor of Massachusetts from 1811 to 1812. During Gerry's governorship, the congressional districts of Massachusetts were redrawn with one district so contorted that it was compared to a salamander: the expression "Gerry-mander" was born. The practise itself is even older, as old as the republic itself.

Not surprisingly, gerrymandering has faced many legal challenges. However, in a stunning denial of reality, the U.S. Supreme Court explicitly ruled that there is nothing wrong with politically motivated gerrymandering. Apparently it is just part of "the ordinary and proper operation of political power". Idiots.

Gerrymandering in America has now reached the level of fine art, shamelessly practised by both Republicans and Democrats. The consequential absurdity of some congressional districts has to be seen to be believed.

How bad could gerrymandering get? To win an election you need to win half of the electorates, plus one extra. And in each electorate you need half of the votes, plus one extra. We'll let you ponder that, but it's clear you don't need even close to half of the voters.

True, the actual practise is not as bad as the theoretical extreme. Still, there is no question that gerrymandering has a huge and detrimental effect on American elections. Americans will probably just have to wait for the Supreme Court to come to its senses. Hahahahaha!

Of course the election on everyone's mind is the hilariously horrid battle between Hillary Clinton and the Donald Trump. To be clear, though gerrymandering thoroughly perverts elections for the United States Congress, presidential elections are not affected. The American president is determined by the Electoral College, a separate example of American absurdity that warrants its own chapter.

What about Australia? Historically, gerrymandering has not been a significant issue in Australian elections, and now there are effective regulations against it. In particular, federal elections are administered by the independent and justly respected Australian Electoral Commission. Australian states, however, are another matter.

If an amateur at gerrymandering, Australia has been an absolute master of the related practise of *malapportionment*: creating electorates of substantially different size in order to give greater weight to votes in certain (typically rural) regions. The practise is remembered in conjunction with the notoriously undemocratic (and corrupt) Bjelke-Petersen government in Queensland, though it probably reached its peak with the "Playmander" of the 1968 South Australian election. The practise is still going strong.

The Australian (and American) Senate is severely affected by malapportionment. Of course, the Senate plays a fundamental role in Australia's federal system and is constitutionally determined, but none of that alters the Senate's fundamentally undemocratic nature.

Malapportionment in the state of Western Australian is worse. It is less pronounced than in the federal Senate but it is significant, serves no good purpose, and is entirely avoidable. The Western Australian electorates vary in population from about 11,000 voters to 26,000 which, special pleading notwithstanding, is a flagrant violation of the principle of one vote, one value. The malapportionment in the Western Australian upper house is much greater, with barely an attempt at justification.

Yes, the American electoral system has been politicised beyond all fairness and meaning. However, as we indicated, Australia is in no position to be overly smug. There is plenty here in our home that could do with some fixing.

Puzzle to ponder

Suppose you have 891 voters to be split into 9 electorates with 99 voters each. What are the minimum numbers required to win the majority of seats.

CHAPTER 30

Tally up the votes that count

D vs. G

With an Australian federal election coming up, we've seriously been pondering for whom to vote. That's been depressing, so let's ponder something else: What about our voting system? Is it fair? Are our individual votes combined to choose a worthy winner? As we shall see, this question is even more depressing: the answer is a resounding and necessary "No"!

In many big elections the population is divided into electorates; for each electorate there are one or more winners, and then those winners are used somehow to determine the overall winner. We won't delve into this, though there are of course many contentious issues with the structure of electorate voting. The point is, there are even more fundamental problems: even when the election consists of choosing a single winner in a single electorate, there is no mathematically satisfactory voting system.

Let us first imagine an election with only two candidates, Dopey and Grumpy. In this case, there is no controversy: whoever gets more votes is the winner. The problems begin when a third candidate enters the race. It may still be the case that one candidate receives more than 50% of the votes, and then we can declare them the winner. But what if no candidate gets that many votes?

The simplest multicandidate system is plurality voting: the candidate with the most votes is the winner. For example, Grumpy would win with 40% of the votes, if Dopey and a third candidate, Sneezy, each received 30%. Though commonly used, the potential drawback of plurality voting is obvious: Dopey and Sneezy may well be like-minded candidates, and so it may be that their combined 60% of supporters would be quite happy with either as winner, whilst detesting Grumpy. And this is not merely a theoretical possibility: it is very likely that George Bush won the 2000 U.S. presidential election only because the anti-Bush votes were split between Al Gore and Ralph Nader. Luckily Bush turned out to be a truly great president, and it all worked out really well.

Is there a better alternative? In the 18th century the Marquis de Condorcet came up with a possibility. Condorcet suggested that we simply take all the candidates two at a time: then the candidate who wins all of their head-to-head races is the winner. For example, with the assumptions above, either Dopey or Sneezy would win 60-40 against Grumpy. So, whoever then won between Dopey and Sneezy would be declared the Condorcet winner.

The Condorcet system is very simple, but Condorcet realised the critical flaw: there may not actually be a Condorcet winner! For example, suppose there are 5 voters, and their preferences are as follows:

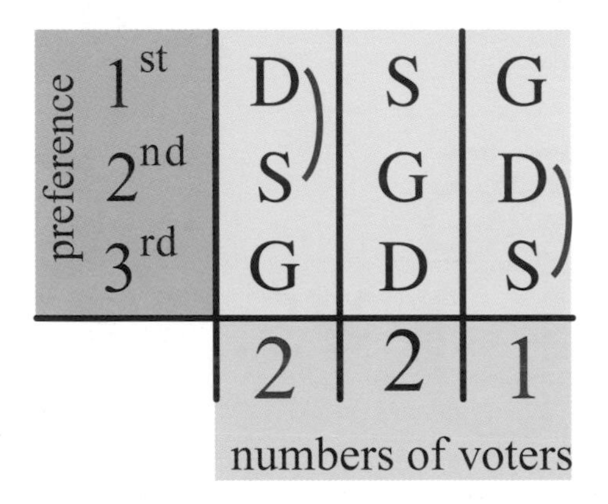

preference			
1st	D	S	G
2nd	S	G	D
3rd	G	D	S
numbers of voters	2	2	1

Then, as indicated in the table, Dopey beats Sneezy 3-2. Similarly, Sneezy beats Grumpy, but then Grumpy beats Dopey! So who is the Condorcet winner? No one!

The Australian method to voting is the *Hare preferential system*. In this system we look at the voters' first preferences, and the candidate with the lowest number of first preferences is eliminated. The eliminated votes are then redistributed to the second preferences, and so on, until only one candidate is left. So, in the previous example, Grumpy is eliminated first, giving one extra vote to Dopey, who therefore wins 3-2 against Sneezy.

The Hare system is fairly natural and always results in a winner (excepting ties), but Condorcet voting shows that this system has its own problems. For example, suppose the two voters in the first column above switch their second and third preferences, giving the following:

preference			
1st	D	S	G
2nd	G	G	D
3rd	S	D	S
numbers of voters	2	2	1

This doesn't change the fact that Dopey is the Hare winner. But notice that Grumpy is now a Condorcet winner! That should make us pause: why should we be happy about declaring Dopey the overall winner if he cannot even win a straight race against Grumpy? But even stranger things can happen. Imagine we had 17 voters, with the following preferences:

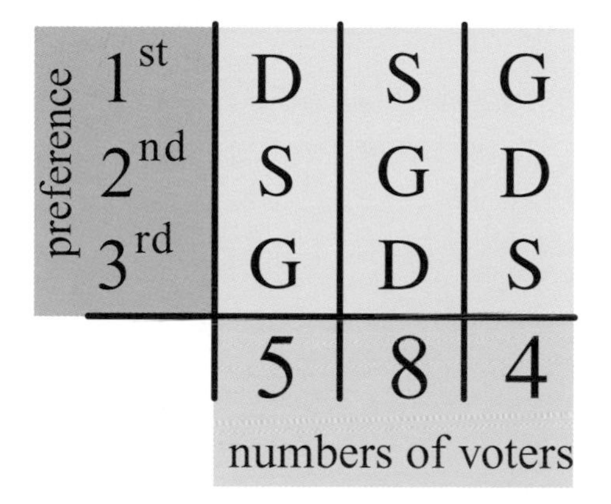

preference			
1st	D	S	G
2nd	S	G	D
3rd	G	D	S
numbers of voters	5	8	4

Then Dopey is the Hare winner, 9-8 over Sneezy. But notice that Sneezy could have been the Hare winner, if only fewer people had voted for him!

Imagine if two of those who voted for Sneezy had instead joined the four who voted for Grumpy. The numbers of votes in the columns would now be 5, 6, and 6. Dopey would immediately be eliminated, and Dopey's second preferences would have ensured Sneezy won 11-6 against Grumpy.

So the Hare system is very far from perfect. Can we do better? Not really. There are subtler systems, which take voters' preferences more accurately into account. But, in 1950, the mathematical economist Kenneth Arrow proved that essentially any voting system will have paradoxes such as the above: there will always be some voting scenario that makes the system look silly.

One obvious question remains: is this all theoretical, or can these voting paradoxes really occur in Australian voting? The answer is, definitely the latter. Here is a possible example, the Senate race in South Australia, in the 2004 federal election.

With five parties left in a tight race, the Australian Democrats were eliminated. The Australian Greens were coming second, but a preference deal between the Democrats and the Family First Party temporarily kept Family First in the race and doomed the Greens. The seat was eventually won by the Australian Labor Party.

Was there anything the Greens could have done? In practise, no. But with hindsight they may easily have won. Similar to the above example, the Greens just needed to have had about 8000 fewer people vote for them! If 8000 Greens voters had instead voted for the Democrats, then the Democrats would have survived and Family First would have been eliminated. The preference deal probably would have next eliminated the Liberal Party, and it is possible that the Greens would have then won the subsequent three-way race. And the moral of this is... well, we don't know what the moral is. Except that voting is weird.

References: There are a number of very good books on voting paradoxes. For example, check out Donald G. Saari, *Decisions and elections: Explaining the unexpected*, Cambridge University Press, Cambridge, 2001.

Puzzle to ponder

There are three voters who will vote for the three candidates R, P, S. Is it possible for the voters' preferences to be such that in pairwise competitions: a) between R and P, then P wins; b) between P and S, then S wins; c) between R and S, then S wins?

CHAPTER 31

Green, with envy

It is comforting to know that Australia has the best voting system in the world. How do we know? Well, there are certainly no visible moves for reform. And, we're sure that this legislative inaction is totally unrelated to the Australian Labor Party and Liberal-National Coalition occupying 98% of seats in the lower house of Parliament.

In the 2007 election the major parties won 85% of the vote, significantly below their subsequent 98% share of seats. It is even more discrepant from the point of view of a minor party: the Australian Greens received about 8% of the vote, for which they were awarded exactly 0% of the seats.

In fact, smug silence notwithstanding, Australia's electoral system is remarkably undemocratic. The system of voting for a single representative in each electorate almost always magnifies the vote of the major parties, resulting in minor parties winning little or no representation.

True, this undemocratic effect is moderated somewhat by the upper house. For the Senate, voting is calculated on a statewide basis, determining twelve representatives for each state. This still favors the major parties, which in 2007 won 80% of Senate votes, garnering them 91% of Senate seats.

However, minor parties and independents have a serious chance of winning fair representation. In 2007, the Greens' 9% of the upper house vote resulted in them winning 8% of the seats up for grabs. And, famously, the Senate can be powerful in "keeping the bastards honest". We'd just prefer there were fewer bastards to keep honest.

Indeed, most European countries have much more proportional electoral systems, similar in effect but more extensive than Australia's Senate system. For example, in the 2009 German election, the Green-Alliance Party received 9% of the vote, for which they received 11% of the parliamentary seats.

True, since proportional systems can lead to parliaments with a very broad spectrum of representation, the outcome is sometimes an unstable and temporary coalition: democracy comes at a price. It also must be said that there are countries with voting systems much, much worse than Australia's. Yes, America, we're looking at you.

Preferential voting in Australia means that a vote for a minor party candidate is not necessarily wasted: as candidates with fewer votes are eliminated, voters' second, third, and later preferences are recorded, until one or more candidates have received sufficient votes to be declared the winner. Neither preferential nor proportional voting is used in the national elections of the U.S. or Britain: in these beacons of democracy, the voting systems can only be regarded as brazen shams.

However, preferential voting is demonstrably not a cure-all. In fact, nothing is. It turns out that, with just the right combination of voters, any voting system will come out looking silly.[1]

To illustrate the paradoxical nature of preferential voting, let's consider a particular electoral race, for the federal seat of Melbourne. In 2007, this traditional Labor seat was won by Lindsay Tanner, with 49.5% of first preference votes. However, with the very popular Tanner retiring, the seat was then being threatened, remarkably, by the Greens.

Based upon the 2007 results, we might guess that the 2010 votes may end up roughly as in the table below. Though Labor would be well in front, the Liberal preferences will flow overwhelmingly to the Greens, and the Greens should win. (This is painting a scenario after distributing the preferential votes of the other minor candidates. In such a peculiar race, this is actually very speculative.)

Labor	40,000
Greens	24,000
Liberal	23,000

And, there's absolutely nothing the Labor supporters can do to save their candidate. Except to vote Liberal.

If a few thousand Labor supporters decide to vote Liberal instead, then the Liberals would receive more votes than the Greens, and the Greens would be eliminated. Then, the Greens preferences would flow strongly to Labor, and Labor would win.

But, of course, such paradoxes are only theoretical. The Labor Party would never actually implement such a strategy. That would simply be a bastardly thing to do.

---

[1]See the previous chapter.

Postscript: As it happened, the Greens did win the federal seat of Melbourne in 2010, and there was nothing Labor could have done about it (except to not be such a second-rate party). The first preferences turned out to be about 34,000 for Labor, 32,000 for the Greens, and 18,000 for the Liberals. So, the Greens polled so strongly that Labor couldn't manipulate the preferences so that they could win.

Puzzles to ponder

1. Suppose that the votes in the federal seat of Melbourne are cast as in the above table, and that Liberal voters always preference the Greens. However, the Greens voters also always preference the Liberals. Can Labor still save itself?
2. Suppose that there are also a number of zombies voting, but not enough to win the seat for the Zombies First Party. However, there are just enough zombies so that, aided by zombie preferences, any of the other three parties might become the winner. How many zombies are there?

CHAPTER 32

Do prime ministers share their birthday cake?

Did you know that Australia's first prime minister, Edmund Barton, and the twenty-fourth prime minister, Paul Keating, were both born on January 18? And, did you know that the American presidents James Polk and Warren Harding share their birthday cake on January 2? Surely, these are amazing coincidences.

What are the chances that both, among the 29 Australian prime ministers and among the 45 American presidents, there are two sharing the same birthday? After all, there are 365 (and a quarter) possible birthdays, and very few leaders to compare. Or, maybe we just hunted for convenient data? What if we had instead checked the days on which our glorious leaders have died?

Probabilities can appear paradoxical. In fact, for any random group of 23 people, there is about a 50% chance that two of them share a birthday. This seems very surprising, but consider the diagram above. When we are asking whether any two prime ministers have the same birthday, we are making as many comparisons as there are lines in the diagram—and there are a lot of lines! In reality, there are 406 lines hiding behind that innocent-looking 29.

Still don't believe us? OK, here's the math. We shall sneak up on the problem: Instead of calculating the chances of a shared birthday, we first calculate the "complementary" chance that there is no shared birthday. Whatever the birthday of Edmund Barton, the probability that the second prime minister has a different birthday must be 364/365. (We've ignored February 29, but including it as a possibility doesn't change the calculations significantly.)

Now that two birthdays are gone, the probability that the third prime minister has again a different birthday is 363/365. And so on. Then, the probability that no two of the 29 prime ministers share the same birthday is the product of all these probabilities:

$$\frac{364}{365} \times \frac{363}{365} \times \frac{362}{365} \times \cdots \times \frac{337}{365}.$$

This amounts to about a 30% chance that no two of our prime ministers have the same birthday. That leaves a 70% chance that two of them will share their cake.

The principle behind the birthday puzzle appears in many unexpected places. For example, imagine you rip 2000 songs onto your mp3 player, and switch it to shuffling songs randomly. If the player really chooses randomly from those 2000 songs, then there is a 50% chance that a song will be repeated among the first 53 chosen.

Puzzles to ponder

1. What are the chances that two of the 150 members of the Australian House of Representatives must share their birthday cake? (An estimate will suffice.)
2. What about the 435 members of the U.S. House of Representatives? (A precise answer is required.)
3. Are there prime ministers or presidents who share their death days?

Part 7

Canned life

Do Americans also have cans? Who would have guessed?

There's obviously nothing particularly Australian about cans. However, somehow we ended up looking at a lot of cans, and especially beer cans. Which is very typically Australian.

CHAPTER 33

The perfect box

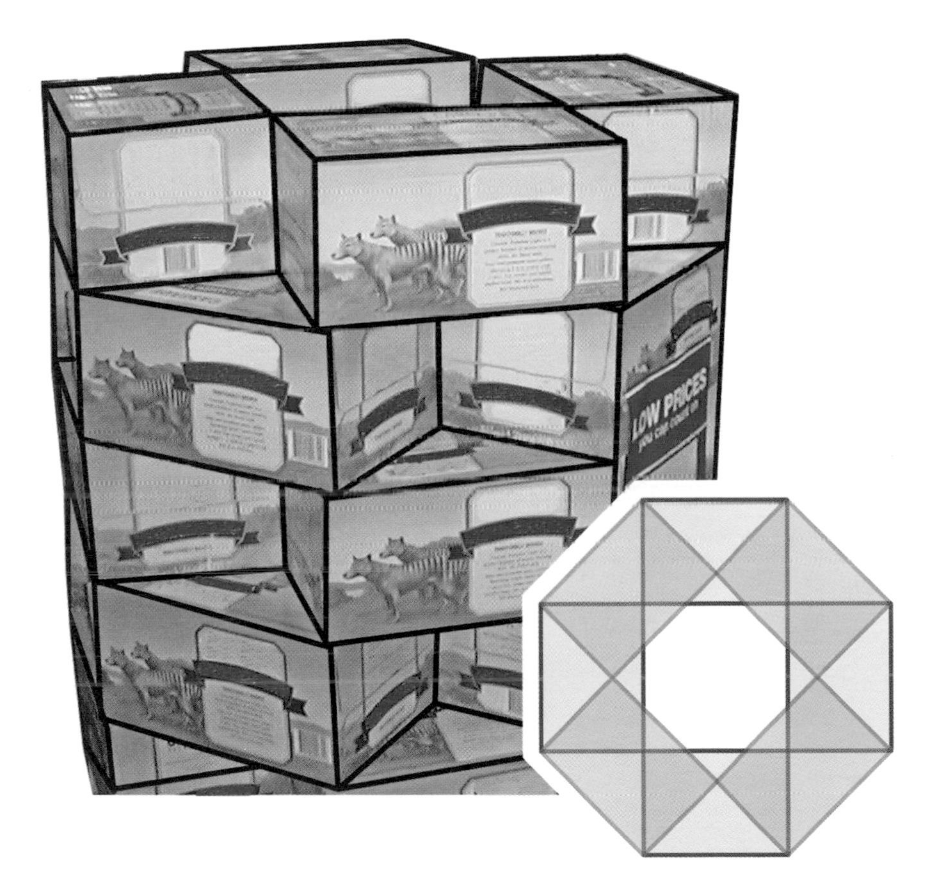

As part of the 2008 National Science Week, the Mathematical Association of Victoria arranged for us to tour rural Victoria. Our mission was to convert kids into math wizards. After a long and thirsty day of converting, we went to a bottle shop, where we discovered a beautiful tower of beer cartons.

It is delightful how neatly the cartons fit together, forming an almost perfect octagonal base. This is only possible because the cartons have rectangular bases in the proportion $\sqrt{2} : 1$. Rectangles of these proportions are famous for another magical property: taking such a rectangle and cutting it in half in the shorter direction, the result is two rectangles with the same proportions.

In fact, you see rectangles like this every day. Standard A4 paper, and all A-size paper, has sides in the proportion $\sqrt{2} : 1$. This paper magic starts with an

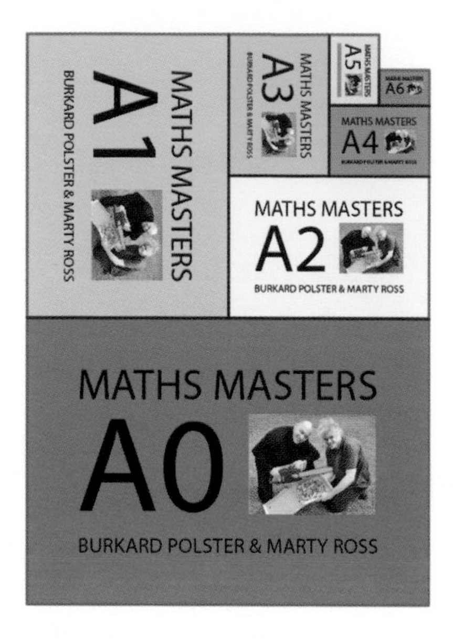

A0 piece of paper, one square meter in area. Cutting A0 paper in half gives A1 paper. Then cutting A1 paper gives A2 paper, and so on.

The advantages of this paper scheme become apparent when photocopying. Placing an A4 document on the copier and hitting the A4 $\to$ A3 button, the result is a perfect A3 copy. Or, placing two A4 pages side-by-side and hitting the A3 $\to$ A4 button, the two pages are perfectly reproduced onto one A4 page.

Alas, there is a hitch in this ingenious scheme: $\sqrt{2}$ is *irrational*. That is, $\sqrt{2}$ cannot be written as the ratio of two whole numbers. The upshot is that an A0 piece of paper should actually have sides of lengths 1189.207115... mm and 840.896415... mm.

It is obviously convenient to use whole number units, and in practise A0 paper is defined to have the dimensions 1189 mm $\times$ 841 mm. So the paper system is not quite perfect, though sufficiently accurate to work well in practise.

But, for mathematicians this slight imperfection is annoying. The irrationality of $\sqrt{2}$ demands the imperfection, and in fact there is long history of irrational numbers annoying mathematicians.

This began more than 2000 years ago with the Pythagoreans. They only believed in whole number proportions, basing their philosophy upon this. So, they were extremely disconcerted when the irrationality of $\sqrt{2}$ popped up in their calculations. In fact, it was only in the 19th century that mathematicians came to fully understand irrational numbers. And still, $\sqrt{2}$ pops up, proving that our paper system doesn't quite work. And still, we're annoyed!

Puzzle to ponder

Arrange three A-size pieces of paper corner to corner into a right-angled triangle.

CHAPTER 34

Crunching can numbers

The other day we purchased the plastic-wrapped slab of beer cans pictured above. Well, no we didn't. The doctored pictured shows $8 \times 5 = 40$ cans. Real slabs contain $6 \times 4 = 24$ cans. But for now, let us pretend: we're fibbing for a greater mathematical good.

Now, being Maths Masters, we noticed that there was an alternative, hexagonal way to pack the cans, within the same plastic wrapper.

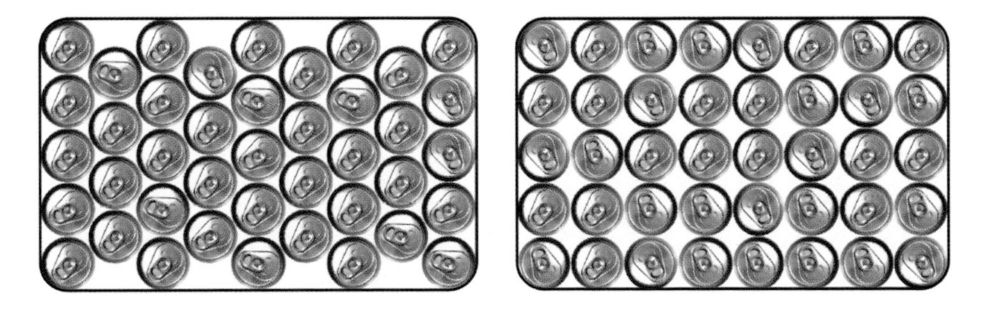

In fact, a fractionally smaller wrapper would suffice. Moreover, this second slab actually contains one more can, for a total of 41.

Why do the beer companies employ the rectangular method of packing? Our brilliant conspiracy theory has them cunningly including larger air pockets to fool consumers on the amount they are purchasing. Once *60 Minutes* reports on our scandalous discovery, the conniving companies will probably turn shame into advantage, by adopting hexagonal packing and thereby "throwing one in for free".

The beer companies may even try to give themselves the mathematical seal of approval. They may point to the fact that hexagonal packing is proven to be the densest way of covering the whole plane.

And they would be right to claim this, as long as we are talking about beer cans with circular cross-sections. But what if we consider cans with other cross-sections? For example, cans with square cross-sections could obviously be packed much more densely, leaving no gaps at all. And the same is true for regular hexagonal and triangular cans.

Obviously, the evil can companies are searching for the opposite extreme. Our spies report that these companies are employing mathematicians to this end. Their task is to find the "evilest" convex (nonindented) shape: the shape whose densest packing is worse than the densest packing of any other shape.

Definitely the circle is not the worst shape in this regard. The strong favourite for this worst cross section is the regular 7-gon. Its densest packing is believed to look as follows.

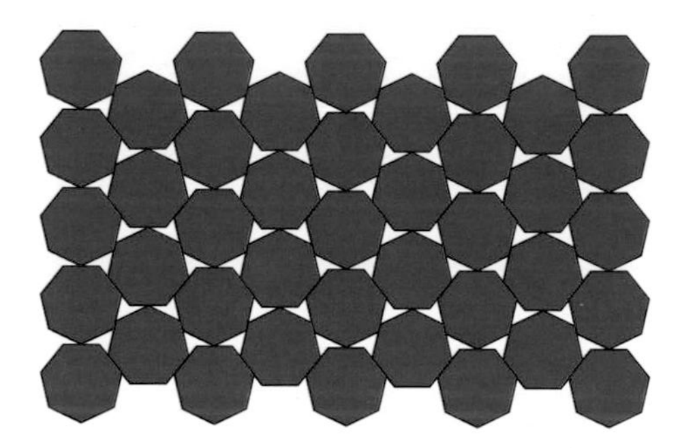

Whereas the densest circular packing contains about 9.3% air, the figure for the pictured 7-gon packing is about 10.8%.

So, will the future be full of 7-gon beer cans? Not if your ever-vigilant, crusading Maths Masters can help it.

Puzzle to ponder

A standard 6×4 slab can also be compared to a hexagonal packing, one containing 24 cans as well. Which of the two packings below has the lower volume?

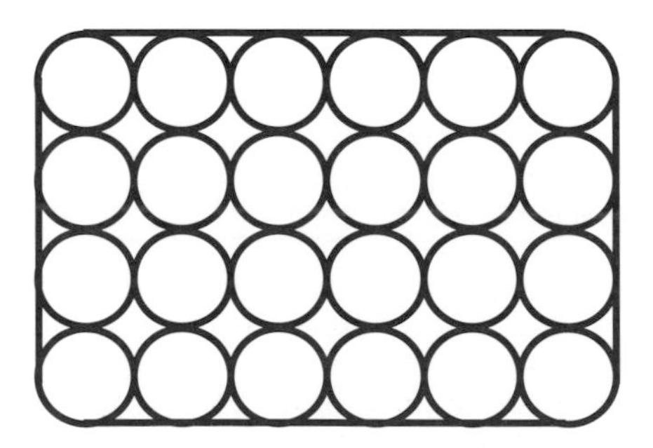

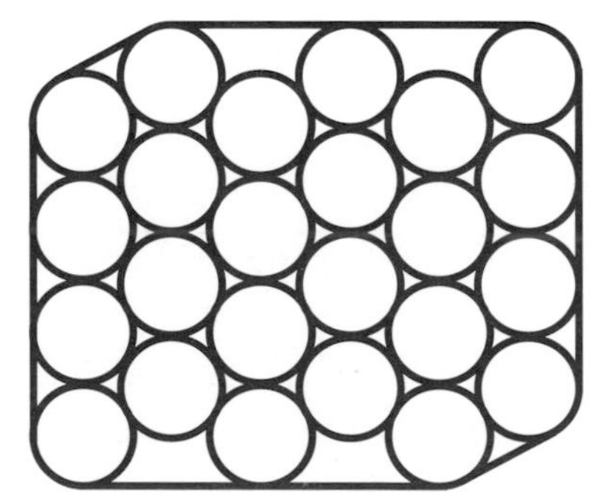

CHAPTER 35

An added dimension to can crunching

In the previous chapter we alerted you to a shocking scandal in the can-packing industry. We noted that the standard rectangular packings included a higher volume ratio of air pockets when compared to the denser hexagonal packings. Today, we report on a true case of soda wizardry.

Have a look at the cubical tower on the left and the hexagonal tower on the right. Each has three layers. If offered one tower to keep, which would you choose?

In fact, it doesn't matter—each tower contains 27 cans. This is no coincidence. No matter how many levels you build into the two towers, one cubical and the other hexagonal, they will always contain the same number of cans.

So, who cares? Probably not shop attendants, who simply want the cans stacked as quickly as possible. But for the mathematically inclined it is a lovely observation. And there is an equally lovely proof.

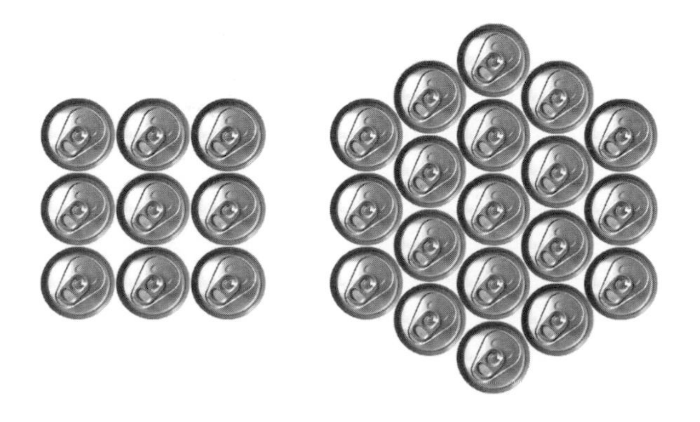

Take another look at the base of the hexagonal tower. Viewed correctly, the outline of a cube jumps out.

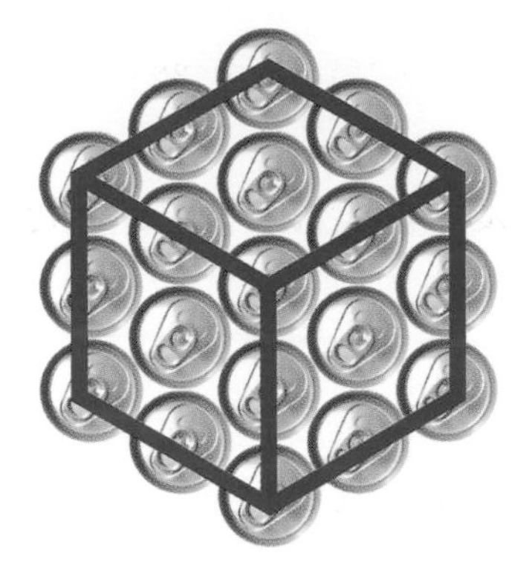

Think of this cube as being made up of $3^3 = 27$ balls. Each of the pictured can tops represent a visible ball. The hidden balls also form a cube of $2^3 = 8$ balls. Therefore, there are $3^3 - 2^3 = 19$ balls that we can see. So, without counting, we can see that the base of the hexagonal tower contains 19 cans. It is also easy to see that there are $7 + 1$ cans in the top two layers, summing to a total of 27 cans.

We can now easily sum the cans in any size hexagonal tower, just by repeatedly adding new bases. For example, for a hexagonal tower with four levels we can use cube-perspective to see that the new base contains $4^3 - 3^3$ cans. We already know that the top three levels consist of 3^3 cans. So, the whole four-storey tower contains $(4^3 - 3^3) + 3^3 = 4^3 = 64$ cans.

Puzzle to ponder

As part of our ponderings, we saw that in cubes made up of balls, the numbers of balls in the shells that make up the cube are special numbers. The same is true for squares made up of circles. What are these special numbers?

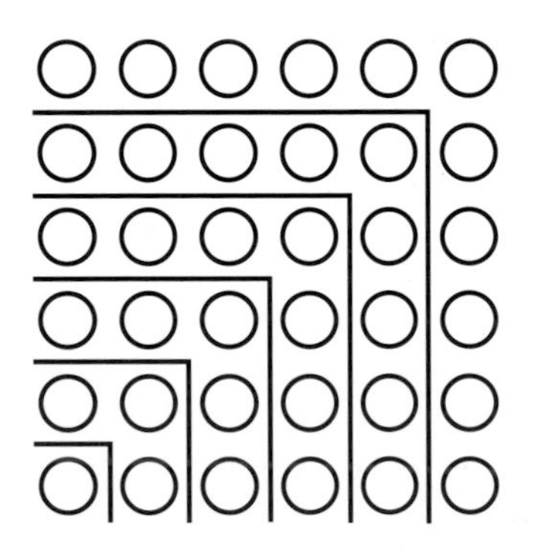

CHAPTER 36

Cool Kepler cat cans

We recently discovered something amazing at our local supermarket: lots and lots of different cans! Yes, we're very easily entertained.

We've written plenty about cans before,[1] but what caught our mathematical eye this time was something quite different. And, this time we're in the good company of the great Johannes Kepler, of planetary fame. As well as deciphering the heavens, Kepler spent considerable time pondering wine barrels, the Renaissance equivalent of our modern cans.

Here's the idea. Suppose a manufacturer is starting a business of selling mouse pâté for pampered cats. She wishes to sell the pâté in 1 kilogram portions (it's for extra-hungry cats) in cylindrical cans. She then has to choose the shape of the can: tall and slim, wide and flat, or something in between.

Of course, even the fussiest cat will care only about the contents. However, the manufacturer will be aware that differently shaped cans have different surface areas. And, the greater the surface area the more metal required, making a more expensive product.

So, what shape cylinder containing a given amount of mouse pâté will have the least surface area? The best cylinder turns out to be exactly as wide as it is tall. That is, in profile it looks exactly like a square. That's a very pretty answer, and it's easy to prove if you know a little calculus.

This then raises the question, Why do we rarely see such cans in supermarkets? Well, there are other factors affecting the cost of production. The pieces of the can are cut from metal sheets, and the circular tops and bottoms cannot be cut without some waste. Also, there are seams where the metal overlaps and is joined. It all gets very messy very quickly.

[1] See the previous three chapters.

There are also factors other than material costs to be considered. For example, it makes sense for a can of soda to be tall and thin, so that it is easy both to hold and to drink from. By contrast, having a flat and wide cat food can makes scooping out the yummy mouse pâté much easier.

Furthermore, it may simply not be worth the trouble to get the shape perfect. For example, pictured below are three vertical cylinders in cross section. They all hold 1 liter (1000 cubic centimeters) in volume, and we have indicated their surface areas (in square centimeters).

The best possible can is in the middle, but we see that surfaces areas of the other cans are not much greater. The difference amounts to less than 4%, which is a cost, but hardly the end of the world.

As for Johannes Kepler, we don't know if he was fond of cats, and it is highly improbable that he consumed much soda. However Kepler did consider a very similar problem involving wine barrels.

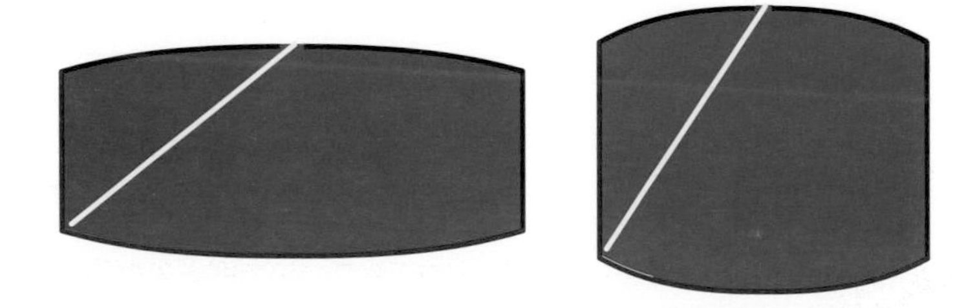

In the 17th century the accepted method for estimating the volume of a wine barrel was to tip the barrel on its side, and then employ a yardstick to measure the distance from the taphole to the opposite edge of the barrel. Using a table of values, different volumes were then ascribed to different lengths.

However, Kepler recognised that it made no sense to use the same volume table for barrels of different shapes. He knew that barrels with the same measured distance could have dramatically different volumes.

Kepler went on to consider the best possible barrel. First, he imagined barrels to be cylinders. Then, among cylindrical barrels with a given yardstick length, he determined the barrel dimensions that would give the maximum possible volume. This was in the days before calculus, so Kepler's calculation was an admirable feat.

It turned out that this maximal cylindrical barrel has a height to width ratio of $\sqrt{2}$ to 1. This means that the profile is the shape of an A4 piece of paper.[2]

Kepler then checked out some Renaissance supermarkets, to see which barrels were actually used. Interestingly, he found that the barrels produced in some regions were very close to the greatest volume for their yardstick measure. It seems that those Renaissance guys just weren't as preoccupied with snazzy packaging.

Puzzle to ponder

How does the diagram below demonstrate that a square has the greatest area among all rectangles with the same diagonal?

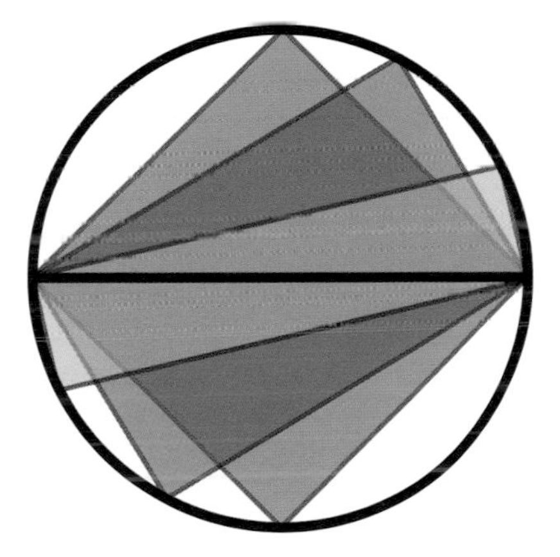

[2]See Chapter 33.

Part 8

Money, money, money

There's nothing quite as annoyingly worthless as an American penny. Australia also used to have 1 cent (and 2 cent) coins but had the good sense to get rid of them decades ago. Of course, the Australian dollar is worth so little, they probably shouldn't have stopped there. Still, if we can't buy much with Australian money, at least we get to play around with it.

CHAPTER 37

Taxing numbers take the law into their own hands

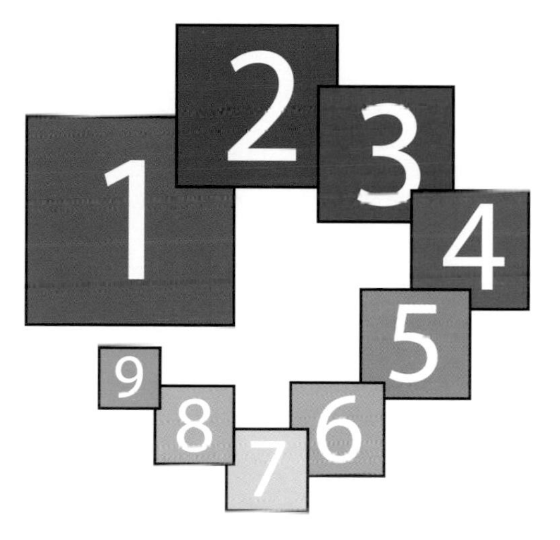

Recently we called the dob-in number[1] of the Australian Tax Office. It took them more than half an hour to answer the phone. Possible explanations for this long waiting time: a) our fellow citizens' new favourite pastime is to dob in each other; b) there is just one person answering the phone; c) ...

Whatever the explanation, it would appear that the tax department is not overly concerned that a dobber may lose interest and hang up. Why? Well, maybe they know about Benford's Law, one of the new mathematical super-weapons used by tax departments throughout the world.

Here is the idea of how Benford's Law is used. An elaborate tax return arrives at the tax department, the data is entered into a computer, and everything adds up. But then they run some fraud detection software over the data, and one check is to determine what percentage of numbers in the tax return start with a 1 in the leftmost spot. The numbers arise from a variety of sources and therefore should be fairly random. So what would you expect the percentage of numbers starting with any given digit to be? Well there are ten digits, so 10% would be a fair guess, right?

Actually, the digit 0 should probably never appear in the leftmost spot. So let's adjust our guess to 100/9=11.1%. Now, here is the surprise. If the tax return has been filled in correctly, then Benford's Law states that the expected percentage

[1] "Dob" is an English word, which roughly translates to "squeal": informing the authorities of some bad behavior, and with some connotation of disrespect for those who cooperate with the authorities.

of numbers starting with a 1 is actually about 30.1%. And, those starting with a 2 is about 17.6%, those with a 3 about 12.5%, etc., down to a mere 4.6% of numbers starting with a 9. It applies to many different statistical data: tax returns, stock prices, population rates, lengths of rivers, and many more. The law was first discovered (not by Benford) more than a hundred years ago, but it is only recently that a satisfying mathematical explanation has been given.

The justification of Benford's Law is complicated, but here is a simple argument which shows why the percentages for each digit should not be equal. If there were equal percentages over all tax data, then these percentages should be *scale invariant*. That is, if the percentages of the digits are equal, then converting from dollars to a new currency should keep these percentages unchanged. But this is impossible. Just imagine a conversion in which you have to multiply all dollar amounts by 2. Then all numbers that started with a 5, 6, 7, 8, or 9 would convert into numbers starting with a 1, and so in the new currency the numbers starting with 1 would now dominate. It follows that equal percentages are impossible!

For a long time Benford's Law was thought of as an interesting but useless curiosity. However, recently Benford's Law has found many interesting applications, pretty much whenever it is worth checking whether data has been tampered with. In the case of our tax return, this means that if the percentages are sufficiently off, the tax man knows to take a closer look.

Reference: *The difficulty of faking data*, by Theodore P. Hill, the mathematician who first came up with a justification of Benford's Law (`digitalcommons.calpoly.edu/rgp_rsr/22`).

Puzzle to ponder

The Benford law percentages for the possible leading digits are

1-30.1%, 2-27.6%, 3-12.5%, 4-9.7%, 5-7.9%, 6-6.7%, 7-5.8%, 8-5.1%, 9-4.6%.

Let's say the leading digits of some tax data are really distributed according to Benford's Law. Now, as in our thought experiment at the end of this chapter, let's multiply all dollar amounts by 2. After this conversion, what percentage of the numbers will start with a 1?

CHAPTER 38

Right on the money for a change

If you're like us, then you always have at least a kilo of loose change rattling around in your pockets. Wouldn't it be a relief to get rid of it all? Here's a novel scheme for doing just that.

Imagine you are at the checkout in a supermarket and the cashier is going through the usual options with you: Reward cards? Cash or credit? Gamble? Hey, that's a new one!

If you decide to "gamble", then you would be asked by the cashier to push a special button on the cash register. For example, suppose your bill is $32.56, and so the closest dollar amounts are $32 and $33. Pushing the Gamble button then produces a random number from 0 to 100, with the exception of 56, with all remaining numbers equally likely. If the number that appears is less than 56, then your bill is rounded up to $33, and otherwise you pay $32. In either case you and the cashier only have to exchange bank notes and dollar coins: none of those pocket-filling silver ones.

If you are more adventurous, you could dispense with all the coins by rounding the bill to $30 or $35. In this scenario a number would be chosen from 0 to 500, with the exception of 256, and you would pay the higher or the smaller amount according to whether your number is lower or higher than 256. Or, you could choose to round to $0 or $100.... However, that may be an overly scary way to buy an ice cream.

The nifty fact about this scheme is that, although you will win or lose on specific gambles, in the long run there won't be any difference to you (or the shops). It all evens out.

We have presented this as gambling and a bit of fun, but it is really just simple applied probabilities. And, in fact, because Australia no longer has any 1 or 2 cent coins, you are gambling like this already. What you pay in a supermarket is not the actual sum of your items, but rather that true sum rounded to the nearest 5 cent multiple. So, each time you pay a little more or a little less, but in the long run it all evens out.

Of course, for our gamble-away-change scheme to function, your decision to push the button would have to be legally binding. Just because \$33 comes up as the amount due, you cannot then decide that you'd rather pay the exact amount. Bad luck!

In fact, with the current 1 cent and 2 cent rounding, you actually do have a choice and effectively can cheat the system. Simply pay by cash if the sum will be rounded down, and otherwise pay the exact amount by credit card. We trust that you'll spend your 1 and 2 cent profits wisely.

Puzzle to ponder

Suppose that after introducing our gambling scheme, supermarkets find out that most people only "gamble" if they are faced with a cents amount greater than 50 cents. What effect does this have on the profit of supermarkets?

CHAPTER 39

Triangulating the Queen

Why are manhole covers round? This question, famously asked at job interviews, has a number of good responses. Our favourite? Covers are round in order to cover up the round holes!

The mathematical answer is less funny, but much more intriguing: manhole covers are round so that they cannot fall through the hole. No matter how you position a circular cover, one of its diameters will be horizontal, and so it will be wider than the (slightly smaller) hole. So, it will always get stuck before falling completely through. By comparison there is definitely a danger with, for example, a square cover. If you align the square with the diagonal of the square hole, it could fall straight through. And the same is clearly true for most shapes.

This leads us to ask, Is it possible to make safe manhole covers of any other shape? The surprising answer is that we can. Above is one solution, a shape known as Reuleaux's triangle. To form this shape, start with an equilateral triangle. Now, draw three circles, each centred at one vertex with radius the sidelength of the triangle. Then the region trapped in the middle is Reuleaux's triangle (see figure below).

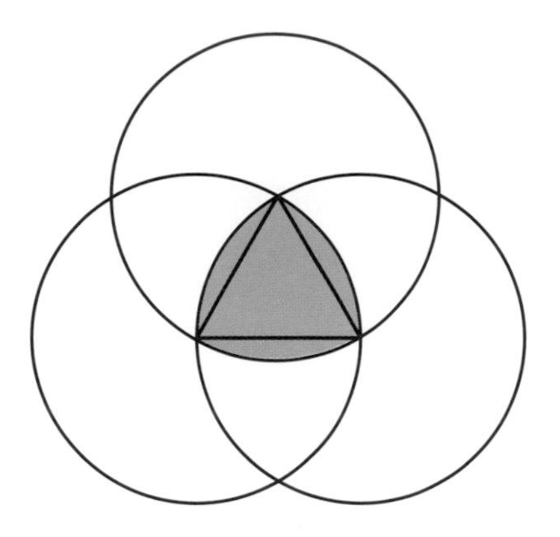

The special property of Reuleaux's triangle is that it is a shape of *constant width*: no matter how Reuleaux's triangle is trapped between two parallel lines, the distance between these lines remains the same; see the picture of the square below. This is the critical property that makes for a safe manhole cover. And there are many other shapes of constant width we could use.

Reuleaux's triangle has other beautiful properties. If you place it inside a square of just the right size, it will roll around smoothly. This is the principle behind an ingenious machine which drills (almost) square holes. And, the famous Wankel engine is based upon a similar principle.

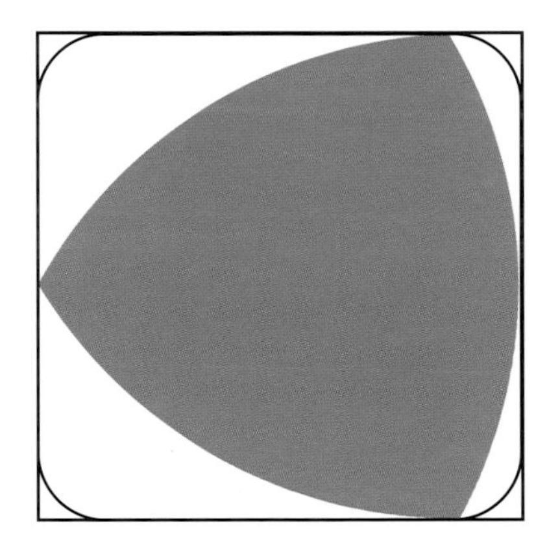

Another property of shapes of constant width is that they make very good coins. A coin machine identifies a circular coin by its width, and this means that the machine could recognise a shape of constant width just as easily. And in fact the British have issued some beautiful constant-width coins.

Which is the best coin? Well, out of all shapes of a given constant width, Reuleaux's triangle has the least area. So, we should make Reuleaux coins, and save all that precious metal!

Puzzle to ponder

Among all shapes of a given constant width, which one has the greatest area?

Part 9

Family life

Sex sells. But, does mathematics about sex sell? Apparently so, since Australian mathematician Clio Cresswell wrote a whole book on sex and mathematics, and it sold very well.[1] We now present our own attempts. We'll see if it works.

---

[1] *Mathematics and Sex*, Allen & Unwin, Crows Nest, N.S.W., Australia, 2004, 192 pages.

CHAPTER 40

Sex, lies, and mathematics

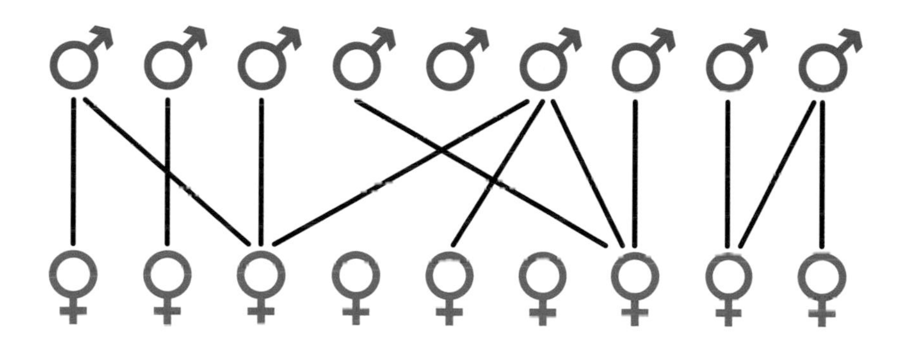

Statistical surveys regularly report that, on average, heterosexual men have more sexual partners than heterosexual women. Consider, for example the survey included in the "Sex in Australia" series, published in 2003 in the *Australian and New Zealand Journal of Public Health*. This was a very large survey, which included 9469 men and 9340 women who identified themselves as heterosexual.

The responses of the heterosexual men indicated an average of 3.9 sexual partners in the previous five years. By comparison, the responses of the heterosexual women indicated an average of only 1.9 partners, less than half that of the men. This difference is amazing. And, for Australia as a whole, it is mathematically impossible.

Let's place all the men and women in two rows and then draw lines between all the "friendly" pairs as shown above. To calculate the average number of partners for the men, we sum the partners for all the men—which amounts to counting the number of connecting lines—and then divide by the number of men. Similarly, we can calculate the average number of partners for each woman.

But of course the number of connecting lines is exactly the same for both calculations. And, as the survey implies, there are about the same number of heterosexual men and heterosexual women in Australia. So the averages must be very close to identical!

The authors of *Sex in Australia* remark upon similar differences in surveys from around the world. A suggested explanation is that women are more accurate in the way they count partners, with (surprise, surprise!) men overestimating their partners. A related factor is the prevalence of sexual double standards: people can simply be too embarrassed (or too macho) to tell the truth.

Of course, such huge discrepancies muddy the true figures, making such reports of limited use. And the message we get is likely to be even less useful, since the media tend only to report these impossible averages.

Math has alerted us to the problem, but how do we solve it? How can we encourage people to tell the truth in such sensitive surveys? Maths may help with a solution as well.

Suppose we are surveying a group, and we want to find out how many within the group were friendly with someone during the previous week. OK, so you ask the question. But, you also tell everybody to make their response dependent on tossing two coins.

The person is instructed to lie if two heads come up and otherwise to tell the truth. Then everyone can tell the embarrassing truth without anyone else being certain whether they are doing so. Similarly, there is now little point in bragging about all your friends. But do these coin-manipulated responses tell us anything? Yes!

Suppose there are 10,000 people honestly using the coin method to respond to our question, with 6000 replying "Yes" and the rest replying "No". Then some simple mathematics shows that there were about 7000 people who were friendly last week. We'll leave you to figure out why.

Reference: As far as we know, the simple argument presented in this chapter is due to David Gale, who used to be a professor of mathematics at the University of California at Berkeley. He is also one of the mathematicians responsible for the *stable marriage theorem*, discussed in the next chapter.

Puzzle to ponder

Explain our estimate of 7,000 friends.

CHAPTER 41

Mathematical matchmaking

Brad	**1. Angelina** **2. Julia** **3. Nicole**	**Angelina**	**1. Brad** **2. George** **3. Johnny**
George	**1. Nicole** **2. Angelina** **3. Julia**	**Julia**	**1. Brad** **2. George** **3. Johnny**
Johnny	**1. Nicole** **2. Angelina** **3. Julia**	**Nicole**	**1. Brad** **2. Johnny** **3. George**

Each year, about 50,000 divorces are granted in Australia. Can all this turmoil and unhappiness be avoided? What if we employed a mathematical matchmaker? Well, that might help...

We shall describe some beautiful mathematics, collectively known as marriage theorems. These theorems show, at least theoretically, how we can achieve the noble goal of matrimonial harmony.

We'll consider the simplest situation. Imagine a Town whose population consists of exactly half women and half men, all of whom know each other. Each man is asked to rank all the women in order of preference, and each woman all the men. (Yes, we can also consider gay marriage theorems.) Then, the job of the mathematical matchmaker is to pair the men and women in a manner that somehow respects these preferences.

Of course, it is not likely that everyone can obtain their first preference. Chances are, the Town will have an equivalent of George Clooney, and only one woman will get him. So, what can we hope to achieve?

Imagine starting with any pairing of the Townsfolk, and then let us ask what might lead to divorce. So, Angelina may prefer the attractive George to her husband Brad, but this would not normally be a problem. However, if it is also the case that George prefers Angelina to his wife Jennifer, then we have a recipe for divorce.

Such a pairing of the Townsfolk, which gives rise to a man and a woman who prefer each other to their assigned partners, is called unstable. And now it is very simple to state a marriage theorem, discovered by the mathematicians David Gale and Lloyd Shapley: *there is always a stable pairing of all the Townsfolk.*

At first glance this is a surprising theorem, but it is actually not too difficult to prove. We first have all the women propose to their favourite men, and each man then provisionally accepts the woman he most prefers. Next, each unengaged woman again proposes to their favourite man (including for consideration men already engaged, but excluding any man who has already rejected them), and again the men accept their most preferred woman; if the man is already engaged, they are permitted to break that engagement to form a new one. This process continues until everyone is paired up.

With a little thought, one can see that this process does indeed result in everyone being paired up, and that the pairing is stable. However, there's a catch: the resulting pairing will not generally be unique. For example, if we allow the men to propose to the women, we will again get a stable pairing, but the pairing will likely be completely different, more amenable to the men. The moral? If your heart is set on George Clooney, don't wait for him to propose to you!

Is this all too reminiscent of Jane Austen? In fact, such "marriages" are now very common. For example, in many countries graduating medical students rank the hospitals at which they hope to do their internship. And, after extensive interviewing, each hospital ranks the applying students. Then, our marriage theorem can be applied to give a stable pairing.

But who does the proposing? In America, it used to be the hospitals. However, after a heated debate, there is now a matching process that favors the students. And in our state of Victoria in Australia? Our matchmaker is the Postgraduate Medical Council of Victoria, but we do not know whom they favor. Demonstrating a coyness worthy of a Jane Austin heroine, they have refused to tell us anything of the methods they employ.

Puzzle to ponder

Try your hand at some matchmaking. The diagram above shows some of Hollywood's finest together with their (made-up) list of preferences. Can you make everybody happy?

CHAPTER 42

Living in the zone

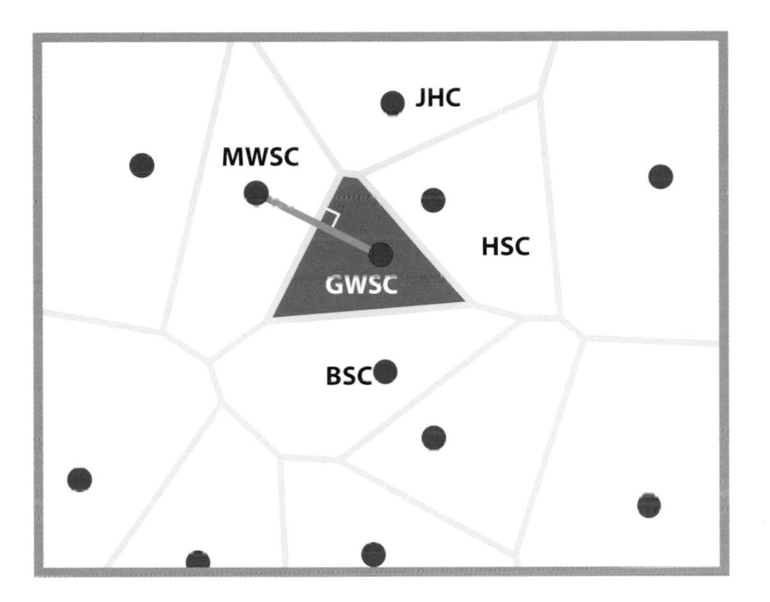

Fortunate art thou who liveth in The Zone. What zone? The neighbourhood school zone of Glen Waverley Secondary School (GWSC) in Eastern Melbourne. Those residing in The Zone are able to attend a government school that, in reputation and results, rivals prestigious, and hideously expensive, private schools.

If you don't live in THE Zone, you still live in *some* zone, for some school. This means that the children in your family must (usually) be accepted by that school. The diagram shows the schools and the zones close to Glen Waverley.

The zone of a given secondary school includes those homes that are closer to this school than any other secondary school (as the crow flies). It is a very geometric definition, with geometric consequences. For example, the border of a zone must consist of straight lines. In fact, joining the two schools in adjacent zones by a straight line, the common boundary of these zones is the perpendicular bisector of this connecting line.

Drawing such perpendicular bisectors, it is easy to construct a complete zone map for the schools in a given area. The end result will always be similar to our diagram. In particular, all zones will be convex (no indentations), which gives the resulting map its characteristic crystalline appearance.

Together, the boundaries of the secondary school zones form a special sort of network called a Voronoi diagram. Voronoi diagrams are useful in many different contexts. For example, a pizza chain can use a diagram to assign a delivery order to the closest branch. More scientifically minded: meteorologists use Voronoi diagrams associated with weather stations to estimate regional rainfall averages, chemists use them to model crystal growth, and zoologists use them to model the territories of animals.

A famous and brilliant application of Voronoi diagrams is due to the British physician John Snow, who solved the mystery of how cholera is spread. In 1854, during the London cholera epidemic, Snow drew a Voronoi diagram associated with the water pumps in the Soho area. Importantly, Snow used travel times to the pumps, rather than straight-line distances. (Shall we stir up a hornet's nest and suggest that Melbourne school zones should be determined by a similar method?) Snow demonstrated that most victims of the epidemic lived closest to the Broad Street water pump. This gave compelling evidence that the water at this pump was itself the source of the cholera.

Voronoi diagrams can also be applied in a reverse manner. Suppose, for example, that you want to live in the Eastern Melbourne region mapped above, but that you really don't like kids, and you wish to be as far as possible from those noisy schools. Where should you buy your house to maximise the distance to the nearest secondary school? The Voronoi diagram shows that you need only consider 26 spots: the 14 corners of the zones, the four corners of the map, and the eight points where the zone edges meet the map edges. That's a lot quicker than checking every house on every street.

This last example may seem a bit far-fetched. But, if you're in a region with a number of toxic dumps, or nuclear sites, or other nasties, a suitable Voronoi diagram could be very desirable.

Puzzle to ponder

What is the Voronoi diagram of the following "hexagonal" arrangement of schools.

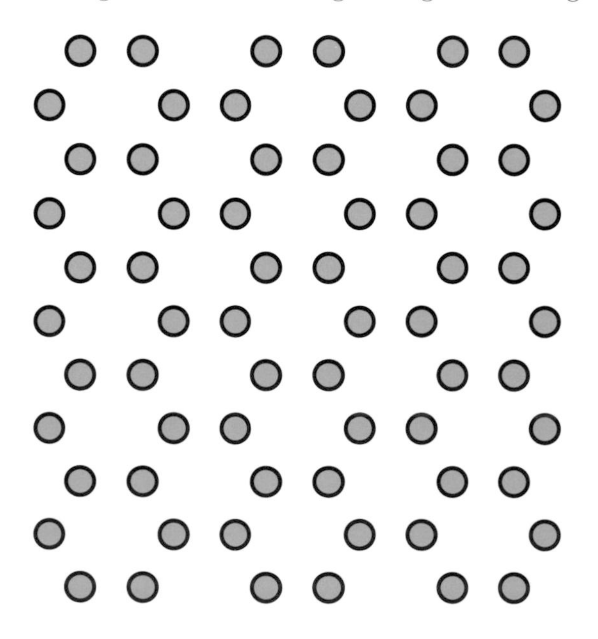

Part 10

Here, there, and everywhere

Australia is very big and, for such a large country, it has a minuscule population. So, long distances and long drives, and speeding tickets, are part of the Australian way of life. And, there's going to the beach, getting caught in the surf, and waiting for a pretty lifesaver to rescue you.

CHAPTER 43

Proving a point on speed

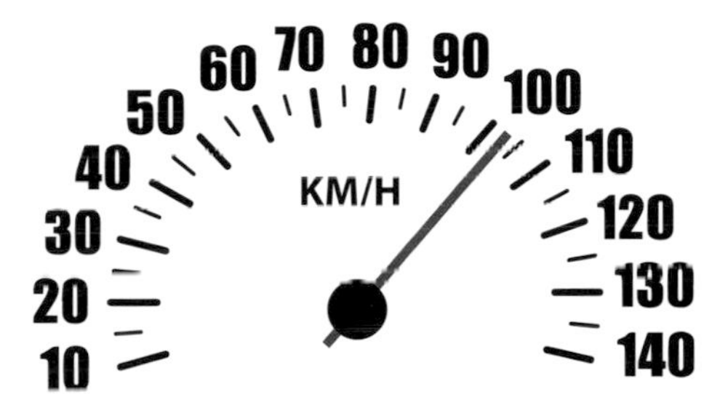

Did you know that speed enforcement dates back to 1903? The problems started in New York State, where incorrigible hoons[1] would speed along at 40 miles per hour—too fast to be caught by the bicycling policemen.

The police responded with an ingenious "point-to-point" system of enforcement. It consisted of three fake tree trunks placed at 1-mile intervals. Each trunk concealed a policeman equipped with a stopwatch and telephone. This setup allowed the first two policemen to calculate the time a car took to travel the mile between their trunks. If the car's average speed was over the limit, the third policeman could then jump out and book the driver.

The system worked, but the drawbacks are obvious. First of all, employing three officers to sit all day in trees may be humorous, but it's also rather wasteful. And, using humans means the system is subject to human error.

However, the point-to-point system is fundamentally simple and reliable. And, on a toll road such as Melbourne's EastLink Motorway, it would be very easy to automate. Each time you pass a toll sensor, a computer registers your license plate and the exact time that it sees you—just like the officers in their tree trunks. And, of course the computer could then do the simple arithmetic to check whether you've been speeding, and then issue your ticket automatically. Such splendid service!

Intriguingly, the computer would not know exactly when the speed limit was broken. The high average guarantees that the driver has been speeding at some moment without actually pinpointing when. This also means the point-to-point system can fail to detect speeding if the driver slows down sufficiently before the next sensor. But why would a speedster bother?

Let's look at an example. Suppose the tollgates are 5 kilometers apart and that the speed limit is 100 kilometers per hour. So, travelling exactly at the speed limit, the trip will take you 1/20 of an hour, or three minutes. Now, suppose instead you decide to speed at 120 kilometers per hour for the first minute. Then you will have travelled two kilometers in this minute, leaving two minutes to travel the

[1] "Hoon" is an Australian word for a lout, and it applies particularly to driving.

remaining three kilometers. That means that for the next two minutes you'll have to be travelling at 90 km/h whilst cars are racing past you. Great fun.

According to the Victorian Department of Justice, there are no firm plans to implement the point-to-point system on EastLink. Instead, motorists can look forward to a traditional battery of fixed radar guns. Unlike the point-to-point system, radar guns can be beaten, if for instance a clever GPS system or app warns you when you are approaching them.

But, after an amusing hiccup, it seems the point-to-point system can again have its day. In 2005 Victoria Police began trialing the system on small stretches of the Hume Highway. This trial missed perhaps the most important point: if the coverage is not comprehensive, then a speedster will be tempted to misbehave along the blind stretches. So, implemented in this way, the point-to-point system has the same drawback as using radar guns. Still, the simple mathematical logic of the point-to-point system is compelling, and the ticketing system went live in 2007. And it was then stopped in 2008.

Amusingly, it turns out the speed cameras were the victims of a software fault, resulting in at least nine (and we're betting many more) incorrectly issued speeding tickets. The moral is, computers are also subject to human error. The cameras were reactivated in 2012 and are now supposedly working well. At least, so says one human.

Puzzle to ponder

Let's say there are three equally spaced toll sensors, 100 kilometers apart. Trucker John passes through the first gate at 2 pm and through the second one at 4 pm (he had a bit of a coffee break along the way). How fast does he have to drive from the second to the third gate to average 100 kilometers per hour from the first to the third gate?

CHAPTER 44

The Maths Masters' Tour de Victoria

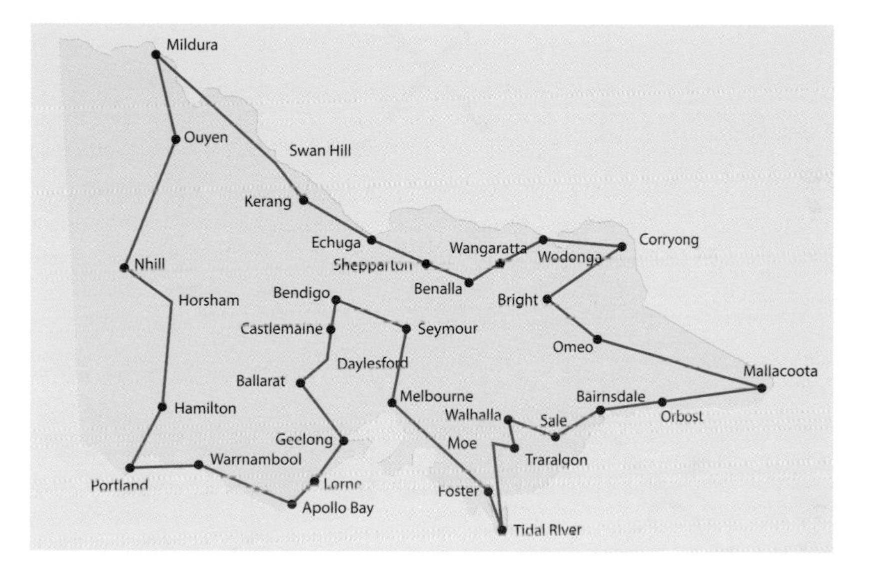

Surprising as it might seem, mathematical speakers are sometimes in very high demand. In anticipation of things going really crazy, we have planned the Ultimate Maths Masters Tour. Our route will take in 34 of Victoria's friendliest and prettiest towns.

The plan is to start in Melbourne and to fly the Mathscopter in straight lines from town to town. Our route will consist of one big loop, finally ending back in Melbourne. There are many loopy routes to choose from, but being Maths Masters we can only be really happy with the shortest possible one.

We are reasonably sure that the shortest tour of Victoria is that indicated in the diagram, coming to a total of 2172 kilometers. Why only reasonably sure? After all, isn't it simply a matter of taking all possible loops and measuring which one is the shortest? Yes. And No.

Starting in Melbourne, there are 34 possible towns for our first stop. There are then 33 possibilities for the second town, then 32, and so on. This means that we have $34 \times 33 \times 32 \times \cdots \times 3 \times 2 \times 1$ possible loops to consider. Oh, but we do get to divide by 2: this monster multiplication counts each loop twice, once for each possible direction of travel. That leaves us with a mere

$$147,616,399,519,802,070,423,809,304,821,760,000,000$$

loops from which to choose.

That's a lot of loops. Imagine we had a billion supercomputers working away, each analyzing a trillion loops per second. Then we'd have all the loops analyzed in about five billion years: probably just in time to see Victoria and the rest of the Earth plummet into the sun, making the whole calculation somewhat redundant.

So, how did we come up with our proposed route? The argument comes in two parts, beginning with a very clever idea. To illustrate, imagine starting with any loopy tour, and choose any two unconnected segments of that tour. They may be anywhere, but to begin with, let's suppose the two segments actually cross.

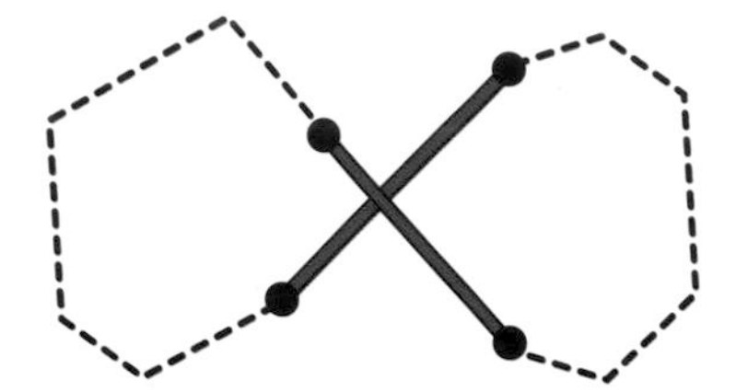

We now consider replacing those two segments with two different segments connecting in pairs the same four towns: there will always be such a choice of segments that also keeps the whole tour as one big loop. And, since the original segments crossed, the new segments will be uncrossed and will definitely shorten the tour.

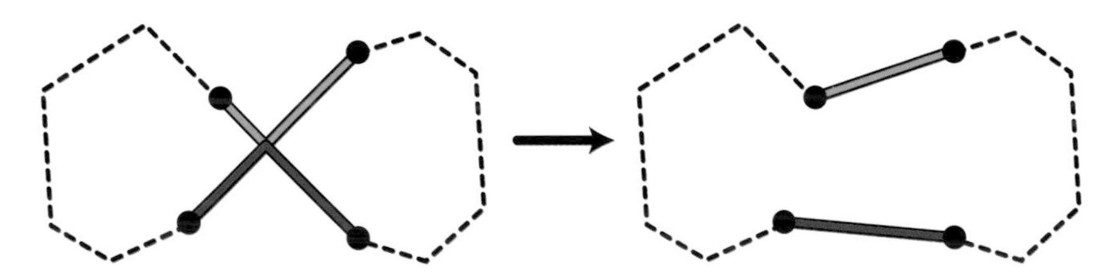

Even if the original segments do not cross, the competing segments may shorten the tour. It's simply a matter of trying and comparing.

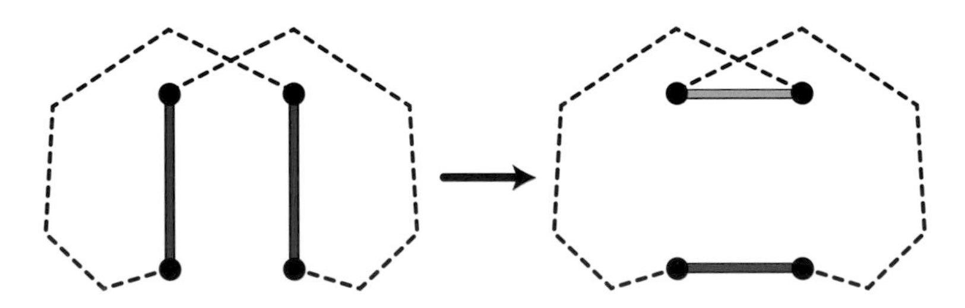

With this simple idea, we now have an easy and relatively quick procedure for coming up with a candidate for the shortest tour. We first start with any loopy route, and we consider all possible pairs of unconnected segments: for our Victorian tour, a total of 561 pairs.

For each pair we then consider whether the suggested interchange shortens our tour. We choose whichever interchange shortens it the most, creating a new loopy tour. Then we try to shorten again, and again, until no interchange shortens further. This final route is our candidate for shortest route overall.

This procedure does not always detect the truly shortest route. However, in practise the approach often gives strikingly good results, very quickly providing a route which is very close to the shortest.

What we have been discussing here is known as the *travelling salesman problem*. The difficulty of this problem, of guaranteeing to have found a truly shortest route, is infamous. No one knows a watertight procedure that does not take an astronomically long time. So, much work is devoted to "reasonably sure" procedures of the type we have described.

Very annoying! Mathematicians hate to say "near enough is good enough". But, when the alternative is to wait billions of years for the answer, sometimes even mathematicians are willing to bite their tongues.

Puzzle to ponder

Let's say there are 64 cities located at the centres of the squares of a chessboard. Find a shortest round trip that includes all cities.

CHAPTER 45

The Ausland paradox

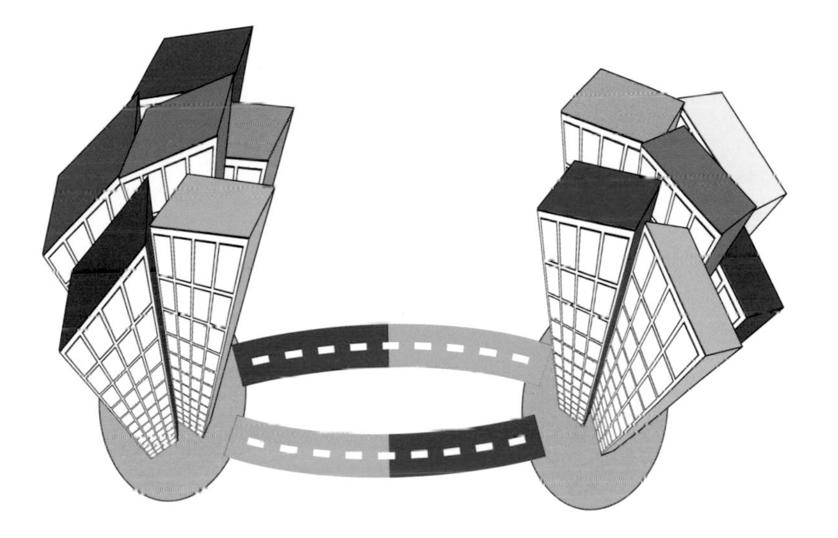

Once upon a time, in the faraway kingdom of Ausland, there were the two mighty cities of Melbaville and Sydtown. The cities were connected by two highways, each consisting of a red section and a green section. A red section always took two hours to traverse, but time on the green sections varied. When traffic was heavy, a green section would also take two hours, but in light traffic an hour sufficed.

Overall, neither highway offered an advantage, and motorists chose between the two routes more or less equally. This resulted in traffic being sufficiently sparse that the green sections were quickly traversed, and the total travel time between cities was three hours. Everyone was destined to drive happily ever after.

But then King Kevin the Klever had an idea.[1] Halfway along, the highways passed through the rustic townships of Canburg and Queanbee. Although very close, no road connected the two towns. King Kevin decreed that a superfast road be built, after which travelling from Canburg to Queanbee took a matter of seconds as shown below.

King Kevin was pleased. For about a day. Then Kevin's loyal subjects began petitioning him, complaining that it now took four hours to travel between Sydtown and Melbaville!

[1] A nod to Kevin Rudd, who was prime minister of Australia in 2009, when this column was written. Kevin wasn't nearly as dumb or as damaging as Tony Abbott, but he did tend to be an out-of-control control freak.

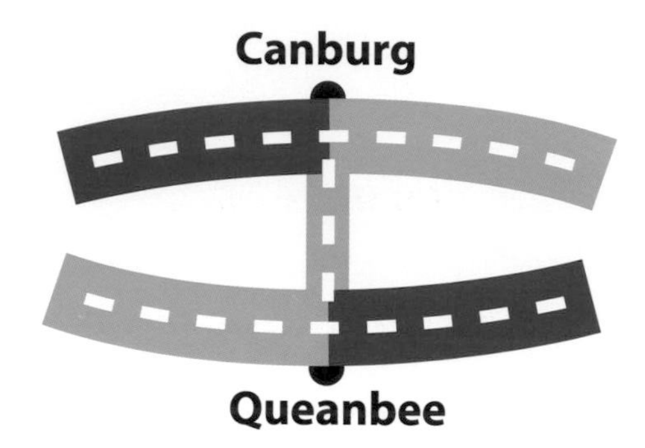

What had happened? King Kevin summoned the royal mathematicians. They explained that the new super-road gave travellers more options. Quite reasonably, most people choose to begin along a green section, switch along the super-road and then traverse the second green section. But, because so many chose this route, traffic was clogged, and the trip now took four hours.

King Kevin was still puzzled, but satisfied. He gave big research grants to the mathematicians to design Ausland's roads. And everyone lived happily ever after, particularly the mathematicians.

That ends the fairy tale, but what about in real life? Obviously, actual road networks are not so simple. Nonetheless, it can genuinely happen that adding a new road slows things down for everybody. And, conversely, closing down roads can result in a speeding up of traffic.

The possibility of such counterintuitive consequences was first predicted in 1969 by the mathematician Dietrich Braess. Subsequently, real-life examples of *Braess's paradox* were observed. For example, road closures in Seoul, Stuttgart, and New York resulted in a speeding-up of traffic. In other cities roads have been identified whose closure is predicted to have the same effect.

We do not know whether Braess's paradox affects cities in Australia. However, there is a similar traffic phenomenon well known to all Melburnians. A fourth lane is being added to the major M1 freeway. Presumably, the hope is that this added lane will speed up traffic. And, it probably will, at least for a while.

But, of course, faster travel time will mean that longer commutes become feasible, which means that more people will buy houses in distant locations, and then add to the traffic. Finally, way too soon, we will have back the same nightmarish traffic. How do we know? Well, lanes have been added to the M1 before....

Is Melbourne's frustrating M1 in the nature of Braess's paradox? Or is it just an instance of a Really Dumb Idea? We'll let the reader decide.

Puzzle to ponder

A truck leaves Melbaville for Sydtown at the same time that a truck leaves Sydtown for Melbaville. They meet at a point 24 kilometers closer to Melbaville than to Sydtown. At this point the drivers switch trucks so that they can return to their hometowns. The driver from Melbaville completed his trip 9 hours after the switch and the driver from Sydtown 16 hours after the switch. If we assume that the trucks maintained constant (but different) speeds, what is the distance from Melbaville to Sydtown?

CHAPTER 46

Schnell! Snell!

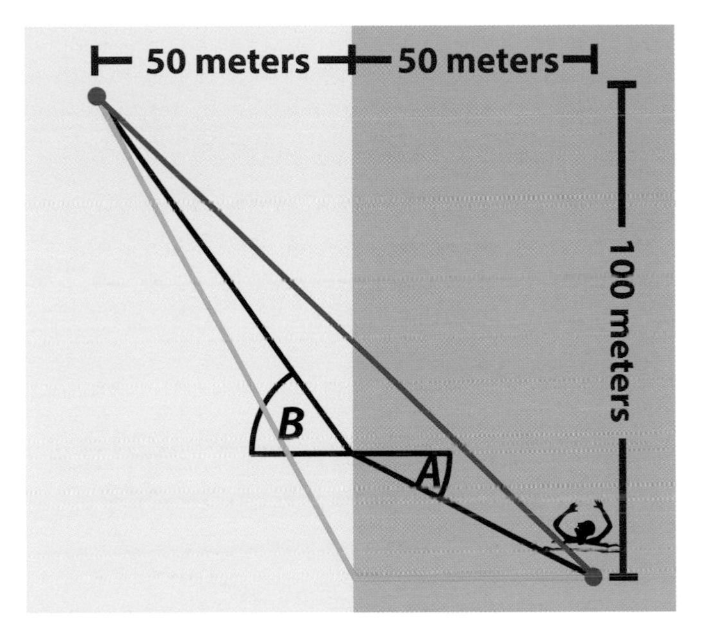

You're relaxing on the beach, and all of a sudden you hear a cry: "Help! Help!" Startled, you drop your holiday maths reading to look out at the ocean. To your right, you see a swimmer, far from shore, waving frantically—to your left on the beach, an athletic lifeguard springs into action. Quick! What is the fastest way for the lifeguard to get to the troubled swimmer? There is no time to lose, and you immediately pick up your pencil and paper, and begin your calculations.

You may (or may not) recall that *Baywatch*'s Pamela Anderson always seemed to choose a straight-line approach (drawn in red). She probably remembered her school geometry, that a straight line gives the shortest distance between two points.

But is the shortest path really the fastest? After all, the lifeguard runs much quicker than she swims. So, perhaps it would be better to minimise the distance to swim, by first running along the beach to the closest point to the swimmer (drawn in green). In real life this is the approach that most lifeguards choose.

Let's consider the situation pictured. We'll assume that our lifeguard runs at a speed of 8 meters per second and swims at a speed of 1.5 meters per second. With the distances indicated, you calculate that it will take her 56 seconds if she runs along the straight red line, and 47.3 seconds if she takes the green route. So, the green route is significantly faster, and very well may be the difference between life and death.

With a deeper analysis, you calculate that the fastest path is somewhere between the extreme paths above. The angles A and B that pin down the bend of this special route are determined by *Snell's Law*, the same law that governs the way light bends when it moves from air to water. Employing our old school friend, the trig function *sine*, we have

$$\frac{\text{speed in water}}{\text{speed on sand}} = \frac{\sin A}{\sin B}.$$

Travelling along this optimal route will have the lifeguard entering the water 8.3 meters earlier, and it will take 46.8 seconds. This gives an extra savings of half a second. After your five minutes of calculation, you are very happy with your conclusion, and you look for the lifeguard to tell her how she should proceed.

Of course, by this time the lifeguard has already dragged the swimmer to shore and is busy resuscitating them. But she'll undoubtedly be pleased to hear that her rule of thumb has the mathematician's seal of approval.

Puzzle to ponder

You are busy building a sand castle when you notice that at the other end of the beach someone managed to set their tent on fire. You grab your trusty bucket and are about to start racing to the ocean to fetch some water when the math puzzling begins again. What is the shortest route from where you are to the water's edge, and then onwards to the tent?

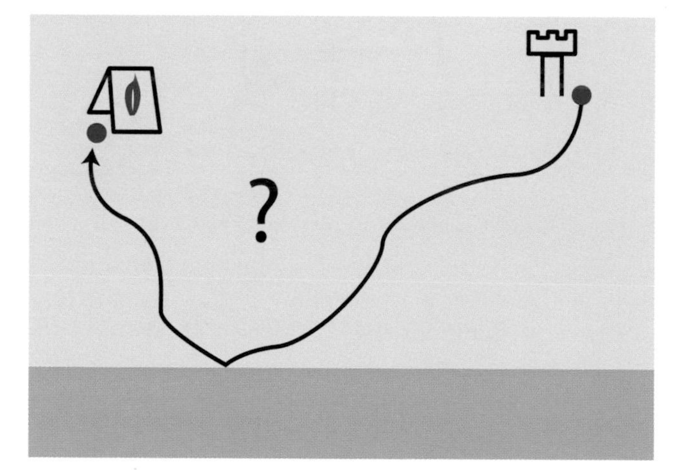

Part 11

TV snacks

Mathematics and mathematicians do occasionally make it to TV. There was the singular show *Numb3rs*, of course, and occasionally mathematicians are the weirdish culprits or victims in murder mysteries. As well, mathematics appears in kids' educational programs, and it's implicit in a lot of TV quiz shows. All in all, there's a wealth of material, and we've been lucky enough to be involved in a few Australian programs. So, are we set to be the newest mathematical superstars? Nope. But we've had a lot of fun.

CHAPTER 47

The Choice is Right!

We have a new idea for a game show. We call it *The Choice is Right!* We hope to sell it because we have a sneaky way of doing better than might be expected.

Here's how it works. The contestant, let's call him Barry Jones Junior,[1] is presented with two suitcases, each containing a number (which may be positive, negative, a fraction, whatever). Barry chooses a suitcase, and then the game host takes the other. If Barry's suitcase contains the higher number he wins a trip to Europe.

Ignoring the chances of a tie, Barry clearly has a 50-50 chance of winning. But now we add a twist. We suppose that Barry is allowed to open his suitcase. Then, before the other suitcase is opened, Barry may choose to swap with the host. Does this opportunity give Barry an extra edge?

It is hard to believe that it could, at least without some more information. For example, if Barry knows that the two numbers are chosen randomly between 0 and 10, then his decision is straightforward: he keeps his number if it is over 5 and otherwise he swaps.

However we're assuming Barry doesn't have any such information. So as it stands, how would Barry decide whether to swap? If his number is 1,000,000, that seems big. But maybe the numbers have been chosen as multiples of a million.

Amazingly, no matter how the numbers are chosen, Barry can give himself an edge. What he does is decide for himself which are the big numbers and which are small. He does this by choosing a third number for comparison, for example

[1]Barry Jones is famous in Australia, first for having been a brilliant game show contestant and then for having been a (less than brilliant) Federal Minister for Science.

according to a standard bell curve. All that matters is that every interval of the number line has some positive chance of containing this third number.

If this third number is smaller than Barry's, then Barry declares his number to be "big" and keeps it; nothing has changed. However if the third number is larger than Barry's, then Barry swaps his number for the host's.

When Barry swaps, there are three scenarios to consider, as indicated in the diagram. If the host's number is smaller than Barry's (in the red zone below), then Barry is actually winning, but he loses when he swaps. On the other hand, if the host's number is in the green zone (between Barry's number and the third number), then a loss is turned into a win by swapping. These two scenarios are equally likely (because Barry's original choice of suitcase was random), and so far, Barry's swapping cancels to have no effect.

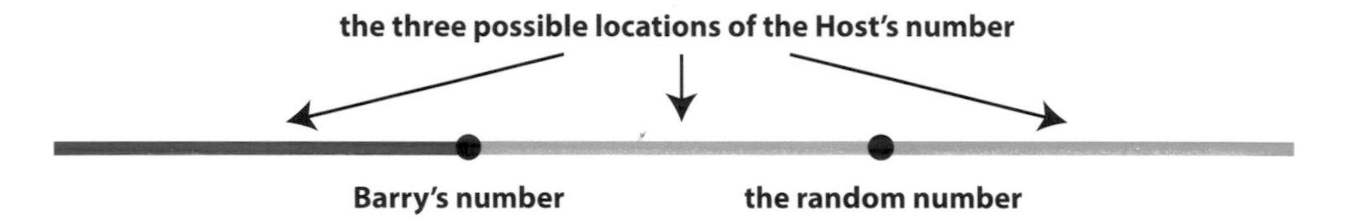

This leaves the blue zone, where the host's number is largest. Here, swapping again turns a loss to a win. And, however the numbers in the suitcases were chosen, there is some positive chance of this third scenario occurring. This is Barry's extra edge!

How much extra edge? That depends upon how the numbers in the suitcases are chosen. If they are also chosen according to Barry's bell curve, then there are six equally likely orderings of the three numbers. Barry's strategy increases the winning orderings by one, and so his odds of winning go from 3/6 to 4/6. And, our plan is that Barry will be so grateful for the extra edge, he'll invite a Maths Master or two along on his trip.

Puzzle to ponder

The Australian version of *The Price is Right* ends with two contestants guessing the value of the final Showcase. They are initially told that the value lies within a \$100 margin, say between \$25,700 and \$25,800. The contestants then take turns guessing (out loud), and they are then told whether their guess is lower or higher than the true value. The contestant to guess the exact amount then gets to play to win the Showcase. What's the best strategy to play this game?

CHAPTER 48

Maths MasterChefs

Cooking shows are all the rage. It seems that every night television chefs compete for culinary honors, and that has given us an idea for another reality show: *Maths MasterChef*.

We want to give our show a trial run. So, here are some of our tastiest mathematical dishes, and you can judge how you think they'll fare. Perhaps they could even add some nutrition to current school offerings.

Entrée: Nourishing Nines Niçoise

For this recipe, take the elusive number $0.999\ldots$, and prepare the equation $M = 0.999\ldots$. This will require infinitely many 9's, so try to buy them on special. Now, gently multiply both sides of the prepared equation by 10. If you were careful, you should find that $10M = 9.999\ldots$. Firmly subtracting the first equation from the second, the result is delectably simple:

$$\begin{array}{rr} & 10\,M = 9.999\ldots \\ - & M = 9.999\ldots \\ \hline & 9\,M = 9.000\ldots \end{array}$$

Now, quickly divide by 9 and you will discover that M is also equal to 1. So, $0.999\ldots = 1$. Very tasty!

Main Course: Traditional Triangle Casserole

To prepare this Greek classic, you will need four red right-angled triangles and a blue square pot.

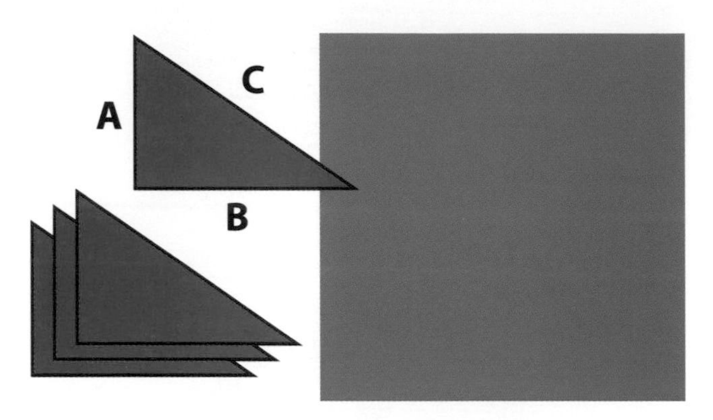

Presentation is everything, so make sure the triangles are identical and are just the right size to fit neatly around the edge of the pot.

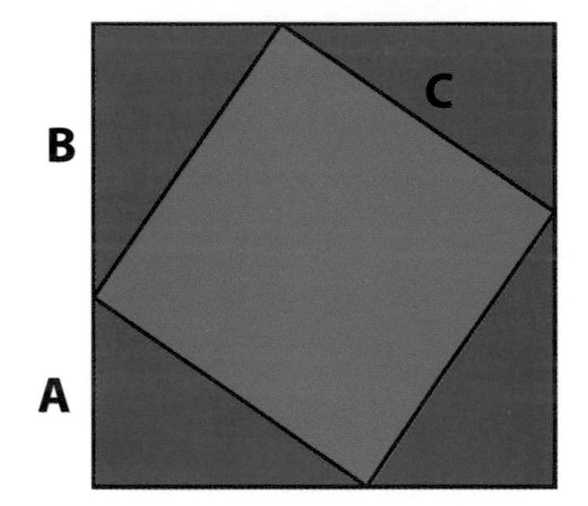

Then the uncovered area of the pot is a square of side length C. That gives an enticing blue region of area C^2. Now, stir the triangles around a bit.

What is our uncovered blue area now? It is C^2, the same as before. Stir again, very carefully...

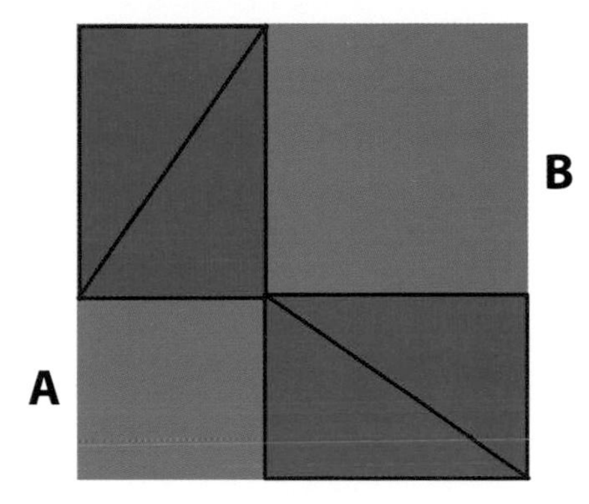

The uncovered blue region still has area C^2. However, the blue region now consists of the two little squares. So, $A^2 + B^2 = C^2$. A wonderfully satisfying dish, a gift of that legendary mathematical gourmet, Pythagoras.

Dessert: π-can Pie

Take an appetizing circle of radius R, and slice it into an even number of identical wedges. The circumference is a delicious $2\pi R$. Now, rearrange the wedges into a parallelogram.

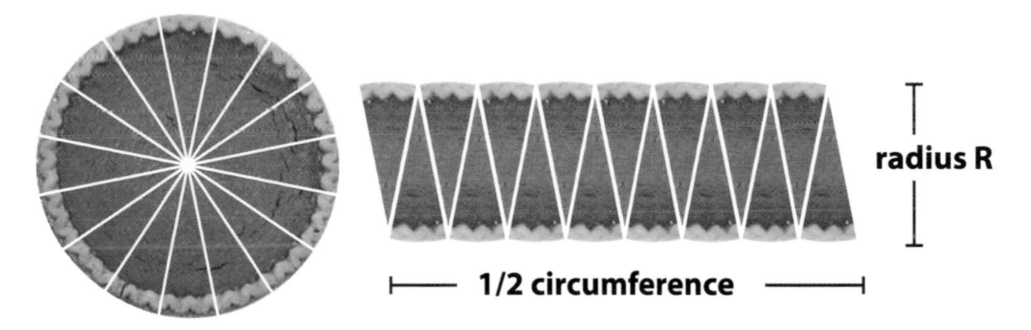

Then the area of the circle is 1/2 of its circumference times its radius R. Very sweet! Combine this with a previously prepared circumference of $2\pi R$, and you can serve up the area of the circle to be πR^2. What a smart dessert to finish a mathematical meal! (A tip for perfectionist chefs: the more wedges and the thinner they are, the more satisfying this dish will be.)

So judges, what do you think? Will such recipes rocket *Maths MasterChef* to the top of the TV ratings?

Puzzle to ponder

And, here is something you can try using leftovers from preparing the Greek dish above. Take two red triangles and place them in a blue rectangular pot. Stir the triangles until you can see at a glance that the blue parallelogram has the same area as a rectangle with the same base and height.

CHAPTER 49

Maths Masters under the Scope

In the previous chapter we jumped shamelessly onto the *MasterChef* bandwagon, sharing some delicious mathematics recipes. When the article first appeared, it somehow resulted in our receiving a call from the producers of *Scope*, an Australian science show for kids. They wanted us to do a segment for their special episode on mathematics. They didn't have to ask twice. *Scope*'s math episode did eventually air, with your Maths Masters featured in the second segment, "Lunch with the Mathematicians".[1]

Though neither of us will win any acting awards, it was great fun, and we got to demonstrate some cool mathematics. However, the segment also moves very quickly. So, for those who prefer to savor their mathematical dishes, we offer here an annotated transcript.

Burkard Polster: Serious and high-powered mathematics. A lot of it is going on in the world. And so is fun and delicious maths, which is the kind we'll be doing today. Hi, I'm Burkard.

Marty Ross: I'm Marty.

BP: We're mathematicians.

MR: Quite hungry mathematicians.

BP: And we'd like you to join us for lunch.

MR: First on the menu is pizza, but it comes with a problem to solve. Burkard is giving me the choice of taking for myself either one large pizza or a medium plus a small one, and I want to choose the option which will give me the most pizza, but I don't know which one that is... until I've done some fast maths to work it out.

[1]You can watch the segment on our website: `www.qedcat.com/media.html`

I cut each pizza in half and arrange the halves so their sides make a triangle. Then I look at the size of the angle between the triangle's small sides. If this angle is greater than a right angle, which is 90 degrees, then the large pizza is bigger than the other two combined. But it turns out that the angle is slightly smaller than a right angle, which means that the small and the medium pizza is my best bet. I'll take these two, Burkard.

The clue for why this works is "right angle". If instead we used square pizzas, and if a right angle is formed, then Pythagoras's theorem ensures that the small-medium combo has the same amount of pizza as the large; and, an angle less than a right angle means that choosing the small-medium combo gives us more pizza. Our trick is based on the fact that Pythagoras's theorem also works if we replace the squares along the edges by semicircles.

BP: All right, after Marty is finished making the most of his smart pizza choice, I am going to present him with a drinks dilemma. I ordered some cordial and water so we could mix our drinks just how we like them. Strong for me, using 2/3 cordial and 1/3 water; and weaker for Marty, 1/3 cordial and 2/3 water.

Now, I'm going to ask Marty to mix the drinks we want, without using any other glasses or jugs. [Marty is scribbling on a napkin.] Well, looks like he's on the right track. Marty pours half of the cordial into the water, and then pours the same amount of that mixture back into the cordial. And he ends up with two equally full glasses, both of them just what we were after.

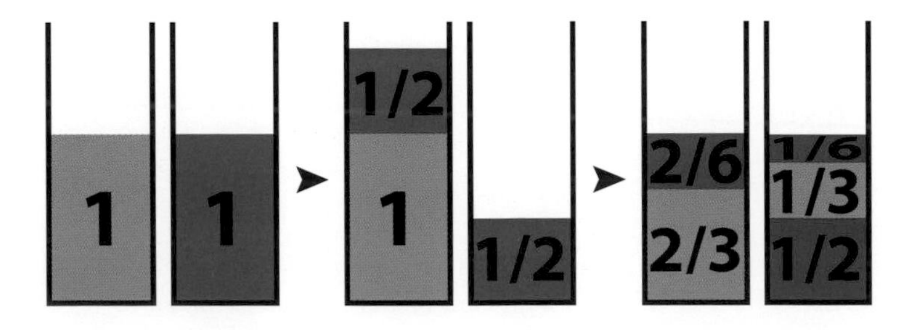

Above is an illustration of how the cordial and water combine. The result is Marty's preferred mixture on the left and Burkard's on the right.

MR: OK, now let's get into some cake maths. What we want to do is to divide this (square!) cake between the five of us so we all get the same amount of cake and the same amount of icing. Cutting the cake like this would give five equal-sized pieces, but some of them would have much more icing than the rest.

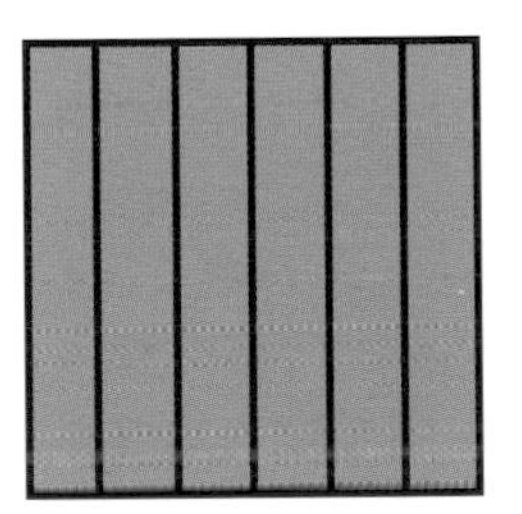

And, cutting the cake like a pizza, making all of the angles the same, will result in unevenly shared icing and cake. So, can maths help us out this time? Of course it can! And all we need is one simple tool. [Marty hands Burkard a ruler.]

Since the cake's edge, the perimeter, is iced, we want to equally share it between the five pieces we cut. So, Burkard measures it (70 centimeters) and divides this figure by 5. Then he uses his answer, which is 14 centimeters, to determine his equally spaced cutting points.

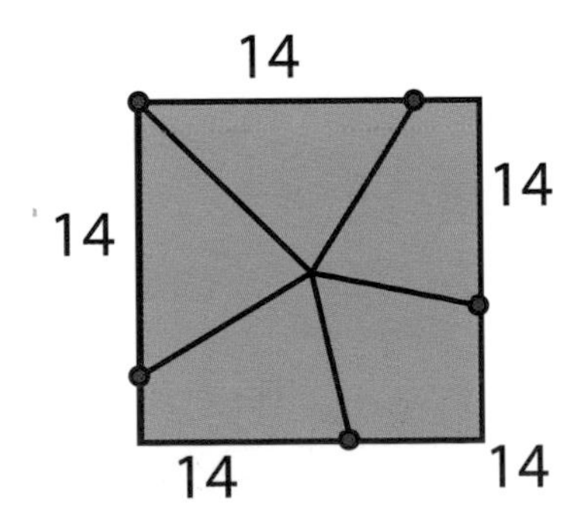

Slicing from these to the cake's centre gives five pieces with equal amounts of cake and icing, which should keep even my most difficult friends happy.

The cuts are designed so that the icing around the edge is divided into five equal parts. However, it is trickier to see that the cake (and so also the icing on top) is also equally divided.

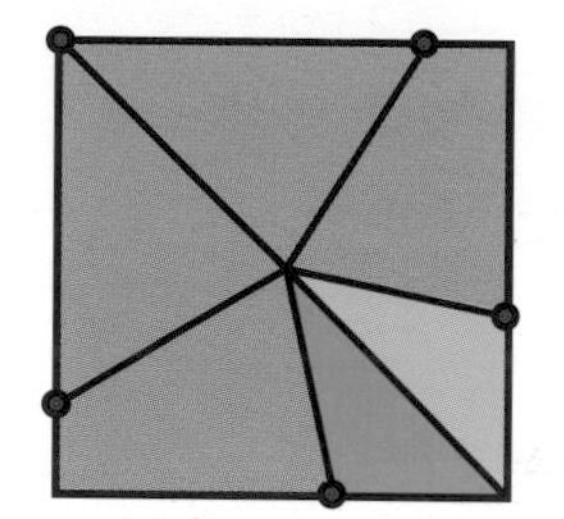

Each slice can be thought of as either one or two triangles with base along the edge of the square: for each slice, the total length of the base(s) is 14 centimeters. As well, all the triangles also have the same height (half the side length of the cake). So, the "half base times height" formula for the area of a triangle guarantees that each slice consists of exactly the same amount of cake (and icing on top).

BP: So, this is how maths can take the guesswork out of serving pizza, cordial, and cake.

MR: Of course, it does work on healthy foods too.

So, that was our mathematical lunch. And all of it packed into a mere three minutes of television. Phew!

Puzzle to ponder

With the same cordial setup as in the text, make Burkard's drink even stronger, 3/4 cordial and 1/4 water, with Marty's weaker, 1/4 cordial and 3/4 water.

CHAPTER 50

How to murder a mathematician

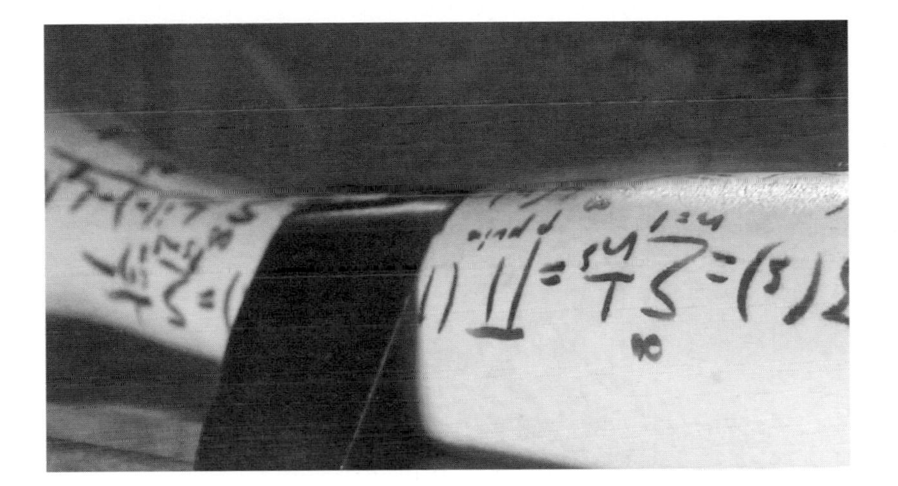

In "Hot House", an episode of Australia's crime show *City Homicide*, two mathematicians were murdered. We helped murder them.

It's not every day that we get asked to participate in a TV drama. And we're actually in the business of collecting and documenting math in the movies and on TV.[1] (It turns out that there's a fairly large number of murderous and murdered mathematicians.) So, when *City Homicide* contacted us to request our help, we jumped at the opportunity.

Initially, we were approached by Kate and Slavko, in charge of props for the show. There were mathematicians to be murdered, and they wanted some help in dressing the sets—providing mathematics books (boxes of them), filling whiteboards with equations, and so on.

In the end, we became quite involved. We checked the math in the script ("Ree-man, not Rye-man"), supplied tons of math props, and spent three days on set, ready to assist with any mathematical emergency. We met the stars, chatted with the director Kate Woods and her crew, and decorated the corpses. All great fun.

The filling of the boards was a lot of work. The first victim was a professor of mathematics, interested in the Riemann Hypothesis, a famously "unsolvable" math problem. So, we dragged out some high-level number theory texts and scribbled

[1]Our book *Math Goes to the Movies* (John Hopkins University Press) appeared in 2012. Our "Mathematical Movie and TV Databases" (www.qedcat.com/moviemath) also contain short summaries of 1000+ movies and TV episodes, with links to hundreds of clips.

away. We included lots of the Riemann Zeta function and related equations, most of which we didn't really understand.

We also needed to fill the whiteboards of a young student. Those were easier and more fun, and we made sure to include as many gems as possible: the golden ratio, the sieve of Eratosthenes, 0.99999 ... = 1, and many more. And, we made sure to include a very prominent QEDcat in the middle of the board. We're delighted that our mascot has now made it to the silver screen.

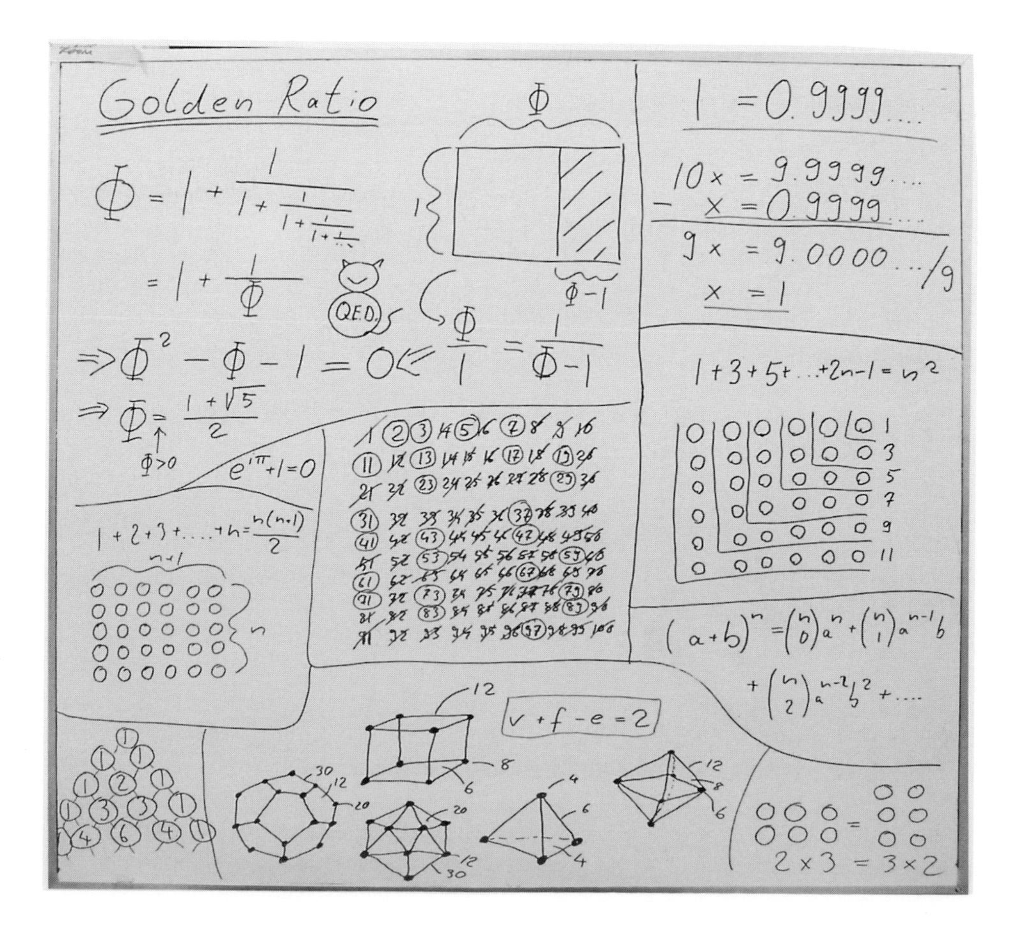

The real fun was in decorating the corpses. The story includes two mathematicians who are forced to write equations on themselves; they get stabbed wherever they make a mistake. We had to write the equations on the bodies, and again we chose mostly Riemann Zeta material. Our two bodies—and Luigi, the practise body—were very friendly, and very tolerant of our crawling over them in order to write the equations.

The whole experience was wonderful. The cast and crew were incredibly professional, and engaging and friendly and fun. If someone ever asks you to kill some mathematicians on TV, our advice is to definitely go for it!

Puzzle to ponder

What other mathematical gems can be spotted on the whiteboard?

CHAPTER 51

A show devoted entirely to numbers (and letters)

A very good quiz show ran in Australia from 2010 to 2012, and it still appears in reruns. *Letters and Numbers* was a local incarnation of the French *Les Chiffres et Des Lettres*, which started in 1965 and is still going strong.

The Australian version wasn't huge in the press, perhaps because the prizes were not exactly humongous: an alternative title for the show might be *Who Wants to Win a Dictionary?* Nonetheless, for fans of Scrabble and arithmetic (and your Maths Masters are both), the show is a lot of fun.

In fact, one of your Maths Masters auditioned to be a contestant. He patiently and fruitlessly waited for his turn, having been officially categorised as back-up cannon fodder. Notably, it was his numbers rather than his letters that let him down.

This was, to say the least, surprising and annoying. What, with all nearby calculators smashed and useless, your Maths Master gets lots of practise with mental arithmetic, and he likes to think he's pretty quick. But, demonstrably not quick enough.

The numbers challenge in the show involves the "small numbers", from 1 to 10, together with the "large numbers" 25, 50, 75 and 100. Six of these numbers are then chosen randomly, and the challenge is to use some or all of these numbers, together with the basic operations of arithmetic ($+$, $-$, $\times$, and $/$), to reach a target number. You are given 30 seconds to do so.

For example, the picture above shows the show's "maths guru", Lily Serna, preparing to use the six numbers 75, 50, 9, 8, 4, 4 to reach the target of 729. One solution for this challenge target is

$$729 = (75 \times 9) + 50 + 4.$$

How does one get good at this? Clearly practise is important, and there are a number of helpful online sites. Indeed, after his humbling audition, your Maths Master decided to practise, and quickly picked up speed. But, it needs more than practise to become a numbers whiz. Strategies are key.

One strategy that often seems to work well is to think in terms of hundreds. Suppose, for example, that our random numbers are the same as Lily's and the target number is 529. That's five hundreds, and we can get very close to our target with another 25. So, we look to make $(20 + 1) \times 25 = 525$. We can actually get this, since $75 - 50 = 25$ and $4 + 8 + 9 = 21$. That gets us 4 off the target, but we have a second 4: a complete solution is then

$$529 = (75 - 50) \times (4 + 8 + 9) + 4.$$

That may seem a little lucky, and you do have to calculate quickly. But with all the arithmetic operations at play, a surprisingly large number of targets can be hit quite easily.

Having practised, your Maths Master then joined the studio audience, to see the show in action. He was reassured to find that he was quicker and more accurate than the contestants (admittedly without the great nervousness of being on national television). But there was also Lily.

Lily can beat your Maths Master hands down. She was clearly much quicker and much more inventive in her solutions. This was really annoying. It was time for Lily and your Maths Master to have a little chat.

Needless to say, Lily is charming. When not being the maths guru on *Letters and Numbers*, Lily was working on her honors research at the University of Technology, Sydney, now completed. Her thesis was on the modelling of pollutant flow in the Great Barrier Reef. It is obviously important work, and it is work that Lily is clearly passionate about.

And what about Lily's skill at the numbers challenge? Surprisingly, Lily said her mental arithmetic was not that strong when the show first came about. However, she practised hard, and she devised a number of different strategies to apply. Lily has become astonishingly quick at applying them.

Lily also indicated one clear trick that really helps: she has learned her 75 times table off by heart. This gives her easy access to numbers where thinking in hundreds will be much clumsier. For example, the target of 729 above would be a struggle for your Maths Master, but is a cinch for Lily.

There were actually some very sharp contestants on *Letters and Numbers*. So, has Lily ever been bested? She paused and smiled: "Rarely".

Puzzles to ponder

1. Using the numbers chosen for Lily above, can you make all the targets $29, 129, 229, 329, \ldots, 929$?
2. Which of the numbers from 1 to 100 can you make using some or all of the numbers, 1, 3, 9, 27, and 81, together with the basic operations of addition and subtraction only (no multiplication or division!).

CHAPTER 52

Bloody Numbers!

NUMB3RS

Numb3rs was eventually axed. After six solid seasons of mathematical sleuthing, the TV series finally came to an end.

We are often asked what we think of *Numb3rs*. People suspect that we're not crazy about it, and that's true. We never watched it regularly and don't miss it much, now that it's gone. However, this has nothing to do with the show's treatment of mathematics.

In fact, though substantially exaggerated in its application, most of the math in *Numb3rs* is solid: they employed excellent mathematical consultants. Our problem with the show is just that we don't think it's much fun. We're still holding out for mathematician characters in a vampire series.

Still, after our experience in getting mathematicians murdered on TV,[1] we became more interested in the math used by forensic investigators. So, we planned to devote the occasional column to some of this gory fun.

We'll kick off with a bloody scene, satisfying our vampire craving, and reminiscent of the *Numb3rs* episode "Magic Show". Our victim has been stabbed, and we want to analyze how his blood has splattered. We hope it will tell us the source of the blood, that is, where the stabbing took place.

What shape are blood splatters? A blood droplet is gooey, and travels though the air as a little sphere, until it hits the floor. This sphere will sweep out a cylinder as it flies. Then, the splatter will roughly be where the cylinder intersects the floor. But then, Euclidean geometry tells us that a tilted cylinder will intersect the floor in an elongated circle: to be precise, an ellipse.

That means the blood splatter should be roughly elliptical in shape. (And, if you're a fan of both ellipses and splattering blood, you'll be pleased to learn that the droplet first contacts the floor at a focal point of the ellipse.)

Our mathematical crime fighter can use this information to reconstruct the path of the blood droplet. The ellipse has two symmetries, and the longer blue axis (see the next page) tells us in which vertical plane the droplet was flying.

The axes of the ellipse make up two sides of the pictured right-angled triangle. So, with a little trigonometry, our crime fighter can easily determine A, the angle at which the droplet has struck the floor.

[1] See Chapter 50.

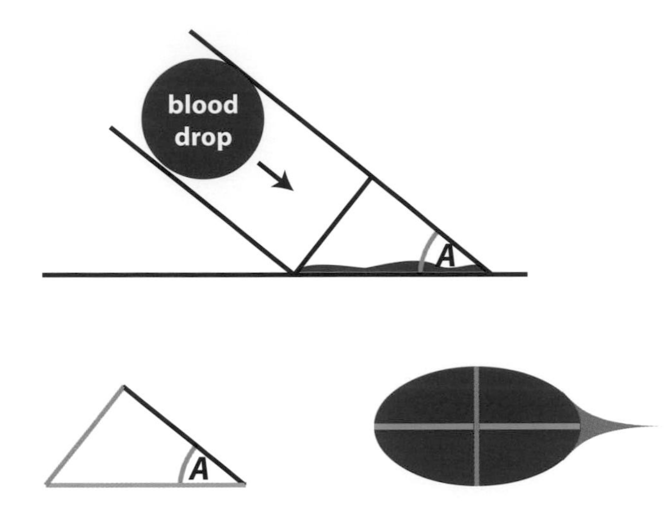

It is also easy to tell whether the droplet flew in from the left or the right. When the droplet hits the floor, smaller droplets will bounce up and leave secondary splatters. Wherever those splatters are, the droplet came from the opposite direction. Finally, combining the trajectories from different splatters, we can compare them to get a very good idea of the source of the blood. Pretty cool!

Of course, this is all very rough, and things are more complicated than we have suggested. For example, the faster a droplet is flying, the more extended its splatter. However, there are more elaborate models to cope with such complications, and it actually does check out: these simple mathematical models work remarkably well. Definitely a great choice for a *Numb3rs* plot. Or, even better, a mathematical vampire plot.

Bloody puzzle to ponder

Two drops of a victim's blood fly from the same wound. The resulting splatters on the horizontal floor are 3 meters apart, and both splatters are 1 centimeter wide and 1.4 centimeters long. Inspecting the long axes of the splatters, and the secondary splatters, you deduce that the drops were flying in opposite directions when they splattered. Roughly how high above the floor was the victim's wound?

Part 12

The Australian math wars

America is currently in the middle of pretty savage math wars. The Common Core has arrived accompanied with all manner of weird teaching techniques and the denigration of traditional skills and algorithms. Prominent math ed professors declare that there's no point in kids learning their multiplication tables. Ugh!

It should be clear whose side we're on. The main mission of "Maths Masters", and much that we do, is to make mathematics fun and engaging, paying little heed to curriculum concerns. But, as do the clear majority of mathematicians, we also value a strong curriculum, with students *properly* learning traditional arithmetic techniques and really *memorizing* fundamental facts.

Does Australia have a math(s) war? No, not really. In America, mathematicians play a significant role in the debate and the formal processes; they may be losing, but they're losing loudly. By contrast, the Australian school curriculum is dominated by math ed clowns, who have little genuine sense of mathematics or understanding of what a mathematics education might offer and what it is really for. For whatever reason, Australian mathematicians have played little role in curriculum debates and have made little public criticism. So, there's no war as such, just a confederacy of dunces in charge, with very few people, such as ourselves, sniping from the sidelines.

We wrote many columns and op-eds on curriculum issues, and we can say with certainty that our writing made not one iota of difference. Except that we really annoyed all the right people. With that, we can be genuinely happy.

CHAPTER 53

Irrational thoughts

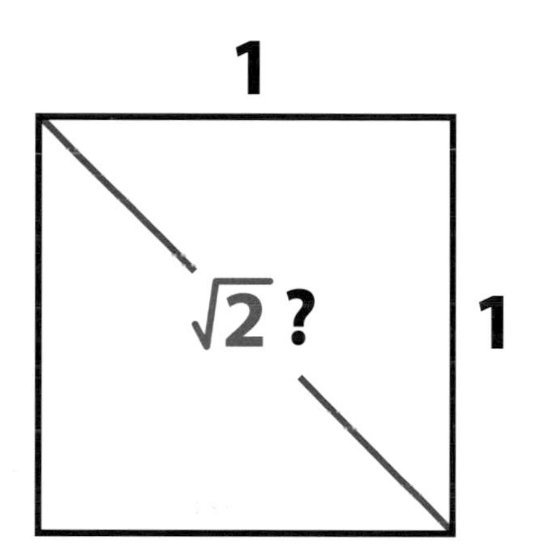

We have real sympathy for the writers of the draft national curriculum.[1]

OK, everybody please stop laughing and let us explain. Consider the following passage from the curriculum, an "elaboration" of Pythagoras's theorem:

> ... recognizing that right-angled triangle calculations may generate results that can be integral, fractional or irrational numbers known as surds ...

True, this passage does not actively encourage sympathy.[2] The expression "right-angled triangle calculations" is hopelessly vague, and what follows is either obvious or wrong.

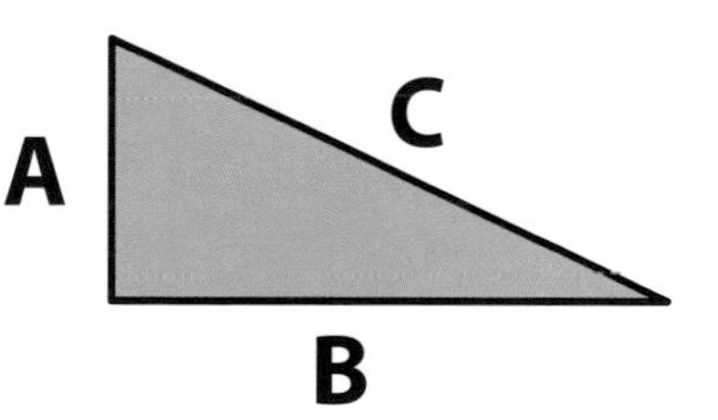

For the right-angled triangle pictured above, Pythagoras tells us that $A^2+B^2 = C^2$. There are whole number solutions to this equation: for example, $A = 3$, $B = 4$, and $C = 5$. It is also obvious that the solutions do not have to be whole numbers: for example, $A = 3\pi$, $B = 4\pi$, and $C = 5\pi$.

---

[1] Australia began developing a national curriculum in 2008, which began to be adopted by the states in 2013. As of this writing, the process is still underway.

[2] If anybody wonders whether we're unfairly harping on an outdated draft, the idiotic passage quoted appears unaltered in the approved Curriculum, as of January 2017.

Interestingly—and this is presumably the point of the quoted passage—the fact that A and B are whole numbers does not guarantee that C is as well. If $A = 1$ and $B = 1$, then it follows that $C^2 = 2$, implying that C cannot be a whole number: $C = 1$ is too small and $C = 2$ is too big.

More mysteriously, if $A = 1$ and $B = 1$, then C also cannot be a fraction. Because of this we say that C is *irrational*, literally *not a ratio*. At this stage, it is common to declare the mystery solved by writing $C = \sqrt{2}$ or $C = 1.414\cdots$.[3]

Not so fast.

First of all, though we may fail to find a fractional solution, how do we *know* that C cannot be a fraction? The question is not even raised in the curriculum.[4] In practise, many textbooks instruct students to push the $\sqrt{}$-button on their calculator, and to then decide whether a number is irrational on the basis of the displayed digits. Such textbooks can be burned without loss.

Indeed, any emphasis upon identifying irrational numbers by peeking at their decimal digits is fundamentally misguided. A finite number of digits never suffices, and it is only for rare, contrived examples that one knows all the digits.

It seems never to be considered that one might *prove* that C is irrational. In fact there are beautiful proofs, completely accessible to junior secondary school students.

Returning to the curriculum quote, if A and B are whole numbers, can C *ever* be a fraction? Putting aside the fact that whole numbers can also be written as fractions, the answer is "No". The curriculum proposition, if it is not boringly obvious, goes awry by suggesting that true fractional solutions are indeed possible.

The above material can be tricky, though it is more a case of people tricking themselves. However, underlying it all is *very* tricky stuff, which has confused much cleverer people. This relates to the last part of the passage: "irrational numbers known as surds".

One question is, What does "surd" mean? Interestingly, "surd" probably doesn't mean anything at all, but that is a story for another day. Instead, let us concentrate upon "irrational".

We have concluded that the solution C to our original problem is not rational, but what is C? Physically or geometrically, we can think of C as the length of a hypotenuse, but to capture the numerical status of C is much more difficult. Writing $C = \sqrt{2}$ is no help, since $\sqrt{2}$ is exactly defined as the (positive) number which multiplied by itself results in 2; as an explanation, as a means to analyze C, this is pointlessly circular. And, $C = 1.414\cdots$ only helps if you can explain the dots. Good luck with that.

It is genuinely difficult to say anything sensible about irrational numbers. The Pythagoreans bumped into them about 2500 years ago, and for most of the time since these numbers have been regarded with suspicion, labeled as "artificial" and "inexplicable" and, well, "irrational". It is only about 150 years ago that irrational numbers were satisfactorily explained. It really is difficult, university-level stuff. And yet, irrational numbers are an unavoidable part of school mathematics.

---

[3] Check out Burkard's *Mathologer* YouTube video, "Root 2 and the deadly marching squares", for a wonderful proof, due to Stanley Tennenbaum, that $\sqrt{2}$ is irrational: `youtu.be/f1yDExNAEMg`

[4] That's changed, fractionally. Pythagoras is introduced in Year 9, and the notion of proving irrationality is introduced, in only the hardest subject, in Year 12.

So, really, we do have sympathy for the curriculum writers. They have made three or four brave attempts to say something sensible about this very thorny topic. It is not entirely their fault that each time they have failed dismally.

Puzzle to ponder

Suppose A, B, and C are the sides of a right-angled triangle. Show that if A and B are whole numbers, then C cannot be a fraction (except if C is also a whole number).

CHAPTER 54

The best laid NAPLAN

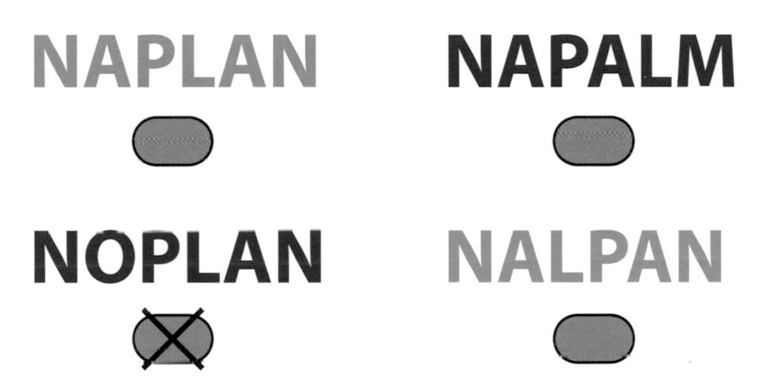

Some weeks are a blast! Of course, every school week is chockablock with fun, but some weeks are special: they're NAPLAN weeks.[1]

Students all over the country in years 3, 5, 7, and 9 will take the NAPLAN tests. They're not exactly fun. Tests tend to be scary and the Australian Curriculum, Assessment and Reporting Authority (ACARA), which is responsible for NAPLAN, assures us that "NAPLAN tests are not tests students can 'prepare' for".

This makes the weeks of preparation undertaken in thousands of schools seem a little silly. It also makes the suggestion that preparing for the tests is somehow cheating seem very silly.

Except, ACARA is clearly fibbing. Any test, unless it is unpredictable to the point of blatant unfairness, can be prepared for. The NAPLAN tests are no different.

In fact, reviewing the tests from previous years, most test questions seem entirely predictable. True, in future years ACARA may bowl a bouncer,[2] setting questions of an entirely different style. If so, we very much look forward to the reaction, and ACARA's subsequent justification.

But who cares anyway? Do the tests actually matter? It's hard to see how a parent, simply concerned for the progress of their child, will learn much of value. Brief numerical results from standardised tests cannot compare to the continuous, close and caring observation of parents and teachers.

No matter how it may sometimes be sold to parents, NAPLAN is not about evaluating students: it's about evaluating teachers and schools. It's carrot and stick, though the stick is big and ill directed, and the carrot is small and distasteful.

Schools are labeled by their NAPLAN results on the misguided My School website. It is planned to award bonuses to teachers on the basis of NAPLAN

[1] NAPLAN is a national testing scheme, which was introduced in 2008.

[2] The cricket version of throwing a curveball.

tests. There are even reports that education ministers have concocted the feather-brained scheme to use NAPLAN results in determining university entrance. Given the current and proposed applications of NAPLAN results, it is very odd to suggest that preparing students for the NAPLAN tests is cheating, rather than just prudent.

Nonetheless, the NAPLAN tests will be well worthwhile if it encourages students to learn. That is, the NAPLAN tests are *only* really worthwhile if they can be prepared for, and if the preparation makes good academic sense. So, it is critical to look at what NAPLAN is actually testing.

Uh-oh.

To begin, the tests exhibit a fetish for jargon. Luckily, your Maths Masters were exempt from the year 5 tests in 2009, where one question required knowledge of a "reflex angle": before looking it up, neither Maths Master had a clue. We were also momentarily thrown by other geometry questions, because words such as "cylinder" have a different, more general, meaning for a mathematician.

In a similar but much more annoying vein, consider the following question from the 2008 year 9 tests:

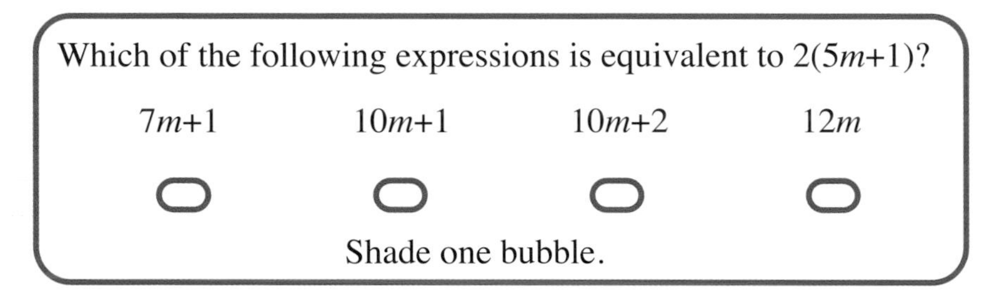
Which of the following expressions is equivalent to $2(5m+1)$?

$7m+1$ $10m+1$ $10m+2$ $12m$

Shade one bubble.

We are tired of people employing the weasel word "equivalent". We wish these testing folk would simply learn to talk of quantities being *equal*.

However, the above examples are largely mathematicians' nitpicks. There are others, but not enough for us to complain too loudly. Unfortunately, there's much more than nits to pick.

The big neon sign of trouble is that NAPLAN claims to test "numeracy", not mathematics. As to what "numeracy" actually means, we're back in the land of jargon: it doesn't mean much of anything. ACARA states that "the main reference for numeracy as well as mathematical knowledge, skills and understanding" is the *Statements of Learning for Mathematics*. The document does not actually contain the word "numeracy". The document is also a clearer demonstration of mathematical ignorance than of mathematical knowledge.

The numeracy game is a cup and balls swindle. While it is commonly claimed that numeracy means something like a basic understanding of mathematics, it is seldom applied that way. In practise, "numeracy" is used to refer to what would be better labeled "functional numeracy", the bare arithmetic and such needed to get by in life.

There is obviously nothing wrong and much right with teaching functional numeracy. We do not object to most of the NAPLAN questions, and some we quite like. We agree that every student should be able to cope easily with the NAPLAN material (except for the stuff on reflex angles).

However, teaching functional numeracy is not the same as teaching mathematics, and it is not a sufficient basis for teaching mathematics. It has little to do with the learning of abstract reasoning, which is the real value in teaching the majority of school mathematics.

True to its title, the NAPLAN numeracy test is not a mathematics test. There is too much concern for placing questions in real-life contexts and not enough testing of basic skills. There are too many decimals and too few fractions, too much estimation and too little calculation. The number π seems not to exist, and of course the word "theorem" is never mentioned, not even old reliable Pythagoras.

Prime numbers appear to have occurred only once,[3] in a very silly question from 2009 asking for the factorisation of 2009. (If it's not clear why this is a silly question, here's a hint: imagine that students in 2011 were asked the corresponding question.)

If it weren't otherwise obvious, the functional numeracy bias of ACARA is made clear in the permitted use of calculators in the year 7 and year 9 tests. This appalling decision is entirely consistent with the techno-fetishism perverting mathematics education in Australia. *Learning for Mathematics*, for example, mentions "theorem" three times while the word "technology" occurs 85 times. The document can be more accurately titled *Learning to Use a Calculator*.

For the teaching of mathematics, and for the basic message of what mathematics is, the NAPLAN numeracy tests are extremely poor. More generally, the NAPLAN tests are a clumsy and misdirected attempt at accountability. Americans are seemingly now coming to the conclusion that, lacking the foundation of clear and good standards, their national testing has been a failure, that it is anathema to a true culture of discovery and learning. When Australians come to a similar conclusion, it won't be a minute too soon.

Puzzles to ponder

1. Factorise 2009 and 2011.
2. Below is a question from the year 5 numeracy test from 2008. What would you answer? What if the skates cost \$38 rather than \$42? Why not just add it up?

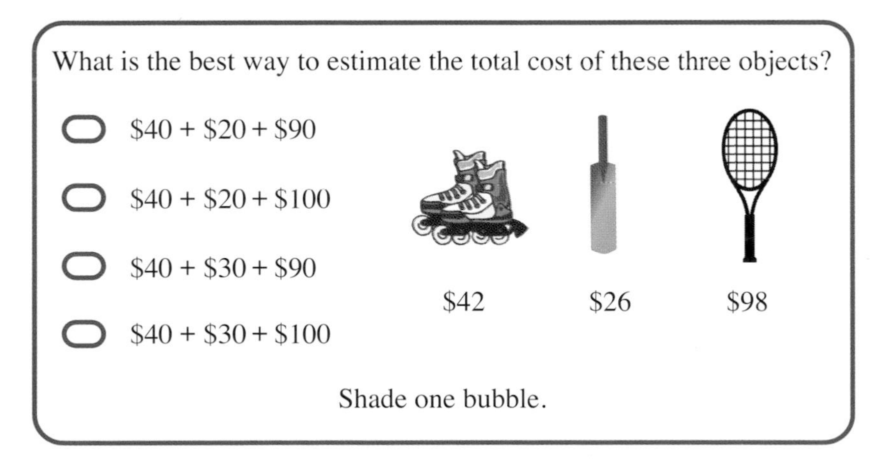

[3] In four years, up to 2011.

CHAPTER 55

Divide and conquer

In 2013 *The New York Times* published an opinion piece on mathematics education. Written by philosopher Alice Crary and mathematician W. Stephen Wilson, the piece could easily be dismissed as just one more missile in America's entertaining but interminable "math wars". However, with a little guidance, we believe that Crary's and Wilson's missile may have struck an interesting and unintended target.

In their piece Crary and Wilson make strong claims for the importance of teaching traditional arithmetical algorithms, the (once) familiar recipes for adding and multiplying, and so on. They argue that applying such algorithms is not, as is often claimed, a matter of thoughtless mechanics. Rather, the mastering of these algorithms requires a sense of the way the algorithms work and a strong familiarity with numbers. Moreover, since the standard algorithms are "the most elegant and powerful for specific operations", it follows that "they are our best representations of connections among mathematical concepts". Crary and Wilson claim that America's "reform strategy" (which has been broadly adopted) trivialises the value of teaching these algorithms and misguidedly encourages the use of calculators in their stead.

Crary's and Wilson's op-ed piece received the predictable cheers and jeers. In particular, it prompted a snarky response from famed math populariser and mathematician Keith Devlin. Though there is much in Devlin's long response, his central objections are easy to summarise.

Devlin notes that the traditional algorithms were invented and perfected for pencil and paper calculation, and consequently they maintain no special status in an era of cheap calculators. Devlin agrees that it is important to teach *some* algorithm for each arithmetic operation, but he claims that there are better algorithms than the traditional ones for the teaching of numerical reasoning. Moreover, Devlin claims that America's reform strategy actively incorporates such preferred algorithms. To the extent that the reform strategy advocates the use of calculators, Devlin has no apparent objection; he specifically claims that having machines perform numerical calculations frees up time to focus upon the deeper nature of number and computation.

Who is right? Who cares?

To begin, Devlin makes an obvious and strong point: just because an algorithm was once fundamental to a mathematical education does not imply that it remains so. Extraction of roots used to be a necessary skill but we would not now think to ask a student to estimate $\sqrt{2}$ by hand. Slide rules and log tables have rightly gone the way of the dodo. Should long division still be taught? Your Maths Masters are undecided but we think perhaps not.[1]

On the other hand, Devlin dramatically undercuts his argument with his claim that calculators can benefit the learning of computation. This claim is *prima facie* nonsense—we have seen no evidence to support it, and we simply don't believe it. We don't believe there is any royal road to understanding arithmetic, that there is any way to learn the nature of arithmetic computation except to compute.

But so what? Your Maths Masters' opinion of Devlin's opinion of Crary's and Wilson's opinion of America's reform strategy can be of little interest to anyone. However, what we believe may be of genuine interest is the light Devlin's response shines upon Australian mathematics education.

Let's start with an easy one: the use of calculators. Crary and Wilson claim, with appropriate contempt, that American reformers advocate calculator use in grade 2. In response, Devlin simply declares their claim to be false. He goes further, professing to not know a single teacher who advocates calculator use in the second grade. We suggest that Devlin get out more, and visit Australia more often.

In Australia it is common for grade 2 classes to use calculators. Moreover, such use has been actively encouraged by education authorities. Indeed, early drafts of the Australian Curriculum had the appalling direction for calculators to be used to introduce counting in foundation classes, to four year olds. Many such blatant and absurd suggestions were howled down and are now out, but the calculators are still very much present. There is no clear directive to *not* use calculators in early grades. By grade 3, and a year before multiplication tables receive their one and only mention, it is expressly encouraged to use "appropriate digital technologies" to solve multiplication problems.

What about algorithms? The word "algorithm" itself does not appear in the Australian Curriculum until year 11. (There, the word appears in a specific and technical sense, unrelated to the issues here.) Can there really be no algorithms at all? It is difficult to tell.

The Australian Curriculum has numerous references to "strategies", but it is often very difficult to determine what is intended. The few arithmetical strategies

---

[1] We've since changed our minds: long division should definitely be taught. And, the approximation of roots, and the teaching of slide rules and log tables all have genuine value. Their roles will never be as central as they once were, but it is important to appreciate the intuition and understanding developed by the application of not-black-box computational devices.

that are spelled out amount to special tricks, such as $57 + 19 = 57 + 20 - 1$. Such tricks are worthwhile but fall far short of the general applicability of an algorithm.

Still, perhaps some algorithms are expected, even if the traditional algorithms are being downplayed. (Devlin did not specify the algorithms he preferred, but we are open minded as to their existence and their value.) However, the simplicity of the sample problems in the Australia Curriculum suggests that algorithms of any type are given little or no weight.

The 2013 NAPLAN Numeracy tests provided further, compelling evidence of the belittling of arithmetic algorithms.[2] The heaviest arithmetic required was probably that in a subtraction problem on the year 5 test: given 1721 of 2350 tickets have been sold, how many are left? (Being a multiple choice question, the problem could also be more easily solved by taking the possible answers and performing addition.) None of the tests appear to have any meaningful testing of multiplication or division skills.

The Australian Curriculum proudly states that "Being numerate involves *more than* the application of routine procedures within the mathematics classroom" (emphasis added). It would then seem fair to conclude that what is supposed to be "routine" in Australian classrooms is to avoid any mental arithmetic that is not trivial.

However, there was one question on the 2013 NAPLAN tests that is definitely worth pondering:

$$37.5 \div \square = 3.75$$

Really? What's to ponder? Not the question itself, of course, which is boring and incredibly easy. What we're pondering is why less than half of year 7 students answered the question correctly, and why only a quarter of year 5 students did so?

Are calculators and the failure to teach algorithms to blame? Your Maths Masters will assume so until a more likely culprit comes along.

So there you have it. Crary and Wilson offer a scathing commentary on America's reform mathematics, which Devlin largely counters by claiming it's attacking straw men, by denying that anyone would advocate such absurdities. And, it seems those same straw men are in charge of Australia's mathematics education.

On occasion, we in Australia watch the goings on in America with a concerned bemusement, but not in this instance. Here, the educational shoe is on the other foot.

Puzzles to ponder

1. Given 1721 of 2350 tickets have been sold, how many are left?
2. What percentage of year 5 Australian students got the correct answer?

[2] See the previous chapter. This column was written in 2013, but the depressing message of the NAPLAN tests is the same every year.

CHAPTER 56

Educational Barbs

Your Maths Masters collect Barbies.

OK, now that we have your undivided attention, we'll get on with the chapter.

In truth your Maths Masters own just one Barbie, however that Barbie is a prized possession. Like many people we've always been appalled by Barbie. Then, in 1992 Barbie's creators outdid themselves by inflicting Teen Talk Barbie upon the world. Each talking Barbie was programmed to repeat four out of hundreds of possible phrases, along the lines of "Will we ever have enough clothes?" Many of the phrases were similarly awful but one was particularly, hilariously awful:

"Math class is tough!"

So, here we are, trying to overcome centuries of the discouragement of women from academic pursuits (and most everything else), and in waltzes Barbie portraying the worst "girls can't do math" stereotype. Of course there were howls of protest, and production of Anti-Math Barbie quickly ceased. She became a collectors' item and is now very rare (and expensive). After years of hunting, your Maths Masters finally obtained one.

It is not only your Maths Masters who have been on the hunt for the sought-after Anti-Math Barbie; she has been a very convenient villain. She makes a guest appearance, for instance, in the recent work of psychologist Dr Sarah Buckley. Dr Buckley's PhD thesis was on the sources of math anxiety, making Barbie a fine symbol for the specific anti-math that Dr Buckley has been analyzing.

Just one problem: Barbie is right and Dr Buckley is wrong.

Well that's not completely accurate. There is much on which we can agree with Dr Buckley and we do not mean to dismiss her research out of hand. However, in promoting her work through various media outlets, Dr Buckley placed particular emphasis upon myths that interfere with a happy and healthy attitude to mathematics. It is the nature of some of these oft-claimed myths with which we take serious issue.

Central to Dr Buckley's discussion are the myths that mathematics (beyond a little arithmetic) is only for smart kids or for kids intending a scientific or technical career. We couldn't agree more; however, Dr Buckley's arguments to debunk those myths are troubling.

For your Maths Masters, the argument that all kids can benefit from learning mathematics is simple: schooling should be primarily about learning to think, and mathematics demands and teaches precise thinking as well as or better than any other discipline. Moreover, we believe that mathematics is beautiful and that everyone is capable of appreciating that beauty. Done.

Dr Buckley briefly mentions the argument that mathematics promotes general thinking skills. However she appears to place much more emphasis upon the familiar and false claim that our modern technological world requires a technologically literate populace. It has somehow escaped everybody's notice that there are millions of technologically illiterate Australians who are nonetheless successful and happy.

Of even greater concern is Dr Buckley's strategy for fighting math anxiety by directly fighting the fear of math and the perception that math is hard. As indicated by the title of one of Dr Buckley's articles, her opinion is, *Relax, there is nothing to fear in mathematics but fear itself.*

Nonsense. There is *plenty* to fear in mathematics.

Great mathematicians struggled for centuries to shape our mathematics into its current form. The 16th-century mathematician Girolamo Cardano referred to the "mental tortures" involved, and the great Gottfried Leibniz discussed his battles with "monsters of the ideal world". These sentiments are typical; mathematics has always caused headaches.

Perhaps it's only more advanced mathematics that is difficult? Arithmetic and other primary school math is pretty easy, right? Sorry, but no. Primary school mathematics is the most difficult of all.

Our modern methods of calculation are based upon positional notation and decimals, and it took thousands and thousands of years to figure them out. Negative numbers? Thousands and thousands of years. Fractions? Let's talk a bit about fractions.

The ancient Greeks, those brilliant geometers, had no proper notion of fractions. Again, it took thousands of years to sort out, and it is still mighty tough. It is so hard to realise that 2/6 does *not* mean "2 divided by 6", that it is not the *process* of division but the end *result* of that process. It is a huge mental leap to realise that 2/6 and 1/3 are one and the same thing, and that thing is only an idea in the mind. It is so easy to get it wrong, and even respected senior textbooks can do so.

Is there anything we can do to make mathematics easier? Of course. We can focus upon the ideas and the time-tested algorithms that simplify the necessary mental processes. We can emphasise those mental processes and the need for sharp mental skills. We can emphasise the beautiful and intuitive patterns, and we can make it all fun.

We can also clear out the needless jungle of jargon, arcane topics, and pseudo-applications. It's worth noting that Barbie didn't specifically complain about mathematics: it was math *class* that she found tough, and she was right to do so. Math class can be more than tough; it can be tough to stay awake.

But the most valuable thing we can do is to simply stop repeating the ridiculous falsehood that mathematics is easy. We can stop replacing genuine mathematics by a cheap fake in an attempt to sell that falsehood. And in particular we can stop pretending that a calculator—that most insidious of lies—can possibly help; we can stop pretending that all that button-pushing is anything more than a cowardly avoidance of mathematics.

Mathematics is tough, pure and simple. Amazingly, for once and only once, we should all sit back for a moment and listen to Barbie.

Puzzle to ponder

Does $(-1)^{\frac{2}{6}}$ equal $(-1)^{\frac{1}{3}}$?

CHAPTER 57

The statistical problem of greedy pigs

Your Maths Masters love games. We're always ready for a sophisticated game of chess, or Go, or Snakes and Ladders. Yep, your Maths Masters are not too old to enjoy games of pure chance, letting the gods of the dice decide our fate. And, we very much enjoy games that require both luck and skill, which makes us big fans of Greedy Pig.

For those unfamiliar with it, Greedy Pig is a simple children's game, played over five rounds. In a given round a die is rolled repeatedly, with each player's total increasing by each number rolled, *unless* the "killer number" 2 appears. To avoid being killed, a player can choose to stop before any roll, locking in their total for that round; the round ends when all players have locked in or when a killer 2 appears, at which point all the greedy players who haven't locked in their scores lose all their points for that round. The winner is the person with the highest score over the five rounds. (Needless to say, there exist many minor variations of the above rules.)

Greedy Pig is a fun game with easy rules. It's also an excellent introduction to the intuition and ideas of probability, making it very attractive as a classroom exercise. So, how does one go about investigating Greedy Pig in a mathematical manner? How does one become Greedy Pig World Champion?

A detailed lesson plan for Greedy Pig is available from the very popular maths300 education website, and a shorter plan in the same style can be found at the NSW Government's Digital Education Revolution (DER) website. As maths300 describes it, the goal is to try to determine the "best strategy" for Greedy Pig. At which point maths300's lesson has already gone somewhat off the rails.

We're not trying to be nitpicky. (If we were, we'd point out to the maths300 writers that the singular of "dice" is "die".) However questions about probability can be very subtle, which means if we're going to be searching for a "best strategy", we really must know what that strategy is supposed to achieve. What is the actual goal? Unfortunately, maths300 never says.

In fact, there is no general "best strategy" for playing the *game* of Greedy Pig. Math can help, but winning Greedy Pig will depend more upon psychology, upon evaluating the strategies of your opponents. For example, if you're playing against a number of greedy players, then playing "sensibly" may result in you beating most of them, but it is very unlikely to make you the overall winner: chances are, a few lucky opponents will choose just the right rolls on which to be greedy. (Exactly the same difficulty arises in trying to win football tipping competitions.)

OK, so if the maths300 lesson isn't about searching for a best strategy to win the game of Greedy Pig, what then is the goal? What more straightforward mathematical goal might there be?

The point (which is implicit in maths300's lesson) is to not consider Greedy Pig as a game against opponents, but rather as a solo game. The individual player is then looking for a strategy to give a good chance of getting a high score. That is, the player wants to try to make her score *on average* as high as possible. This is now a precise mathematical goal, and we can get to work searching for a best mathematical strategy to achieve that goal.

As a lesson, Greedy Pig is intended to be very open and exploratory. Students are encouraged to come up with their own strategies, games are played, and outcomes are compared. Then, hopefully, there is extended discussion of why some strategies may be better than others. Why, for example, is "stop after three rolls" not a great strategy?

All this seems natural and good. The lesson is fun and it may well result in some students gaining some genuine understanding of the way probability works, without resorting to formulas or heavy calculation. If students simply learn that dice don't have memories, that every roll provides the same $1/6$ chance of killing everyone and no matter how long since the last 2, that would be great. It would put the students well ahead of the thousands of foolish victims blowing their savings on the pokies.[1] Unfortunately, maths300 (and DER) goes seriously off the rails.

Exploration and experimentation is all well and good, but so are actual answers. So, what is the actual best strategy for averaging a high score on Greedy Pig? How can we determine that strategy?

After all the games, maths300 concludes its lesson with a computer. They decide to "[use] a computer to find the optimal strategy". The various strategies are programmed into the computer, some simulations are run, and the winning strategy is declared.

Oh dear.

Let's put maths300's computer aside for a moment and get back to actually thinking about the problem. Suppose you happen to have scored 15 points on a round, and you want to decide whether it's a good strategy to continue. Well, there's a $1/6$ chance that on the next roll your score will go up by 1, and a $1/6$ chance your score will go up by 3, and so on; *and*, there's a $1/6$ chance you'll be hit by the killer 2, meaning your score will go down by 15. That means that on average your total on the next roll will change by

[1] See Chapter 27.

$$\frac{1}{6}(1+3+4+5+6-15)$$

Since the above sum is positive, the conclusion is that if you've only obtained 15 points, then (on average) it's worth risking another roll. And what if, more generally, you've already scored a total of T points? Then, on average, the next roll will change your total by

$$\frac{1}{6}(1+3+4+5+6-T)$$

All that matters is whether this sum is positive or negative. The conclusion is that you should keep going until your total on the round is 19 or above, and then stop.

Seriously, how hard was that? How can maths300 provide 15 pages of lesson plan without a single mention of this simple and conclusive arithmetic approach?

There seem to be two underlying reasons for maths300's approach, neither all that convincing. The first reason involves the intended audience: the maths300 lesson is supposedly designed for classes from year 3 to year 12, and "attempted explanations involving averaging and long-run frequencies call upon sophisticated concepts that the student may not yet possess".

Fair enough, up to a point. One would of course wish to be gentle with year 3 students, and the formulas above would be too much. However at some level, *well* below year 12, the simple arithmetic just has to be presented. Indeed, the T formula could be a very natural and impressive introduction to the power of algebra.

Moreover, though the notion of long-term average may be (somewhat) sophisticated, it is also intrinsic to an understanding of the problem, of what the "best strategy" is supposed to achieve. Shying away too much from such ideas makes the whole exercise meaningless.

Bizarrely, maths300 is even reluctant to mention any notion of probability, that the killer 2 has a $1/6$ chance of occurring on each roll. Supposedly such an approach "can be mystifying for a student who self-evidently has seen that the 2 sometimes occurs twice in a row and at other times might not occur for 20 rolls". Well, yes, it can be mystifying. It's also exactly the kind of mystification that a classroom lesson on probability and randomness should be confronting head-on.

The second reason underlying maths300's approach is even more concerning: a fundamental hostility to theory and the very notion of proof. The maths300 writers contrast "empirical versus theoretical" methods of learning and teaching, and clearly have no respect for the latter. For maths300 it seems that mathematics, whether in year 3 or year 12, is nothing more than an experimental science. The notion that a clear statement of probability or a simple formula for an average may be valuable or may solve the problem in a clean and conclusive manner, seems to carry no weight. (maths300 does offer a "theoretical approach" in an afterthought "Answers" document; however, the probability calculations there include none of the simple approach we've outlined above and pretty much answer nothing of interest. It also raises the further question of whether the writers know or care about the difference between a probability and an average.)

Your Maths Masters definitely appreciate the role of computer simulation in teaching Greedy Pig. The repeated rolling of a die can only provide intuition up to a point; after that, to see what "long-term" is really like, one has no choice but to trust the computer and see what happens with zillions of simulated rolls. Moreover, a game needn't be much more complicated than Greedy Pig so that computer simulation is the *only* way to get a reasonable understanding of the probabilities involved.

However Greedy Pig *is* a simple game, and it *does* have a simple, and simple to prove, solution. It is absurd to not present this solution to any student, at any level, who has some chance of understanding it.

Your Maths Masters have heard good things about maths300, that many of their lessons are well worthwhile. Maybe, maybe not. However, in the case of Greedy Pig, it appears that a dubious educational ideology has won out over simplicity and mathematical common sense.

Postscript: This column was pretty much the final straw for us, convincing us that Australian mathematics education was beyond repair. The column received (by our little column's standards) a large amount of heated criticism, none of it worth a dime. None of the many maths300 fans could face up to the fundamental absurdity of the way the game had been presented. Then, in 2016, the writers of maths300 contacted us, indicating they had altered their Greedy Pig lesson in response to our criticisms. However the lesson remained (and last we checked was still) a constructivist swampland, just with the formula solution thrown somewhere in the middle.

Puzzle to ponder

Suppose the rules of Greedy Pig are changed so that on your second roll you get twice the total on the die (except for the killer 2), on the third roll you get four times the total, on the fourth roll 8 times the total, then 16 times, and so on. When should you stop?

CHAPTER 58

The paradox of Australian mathematics education

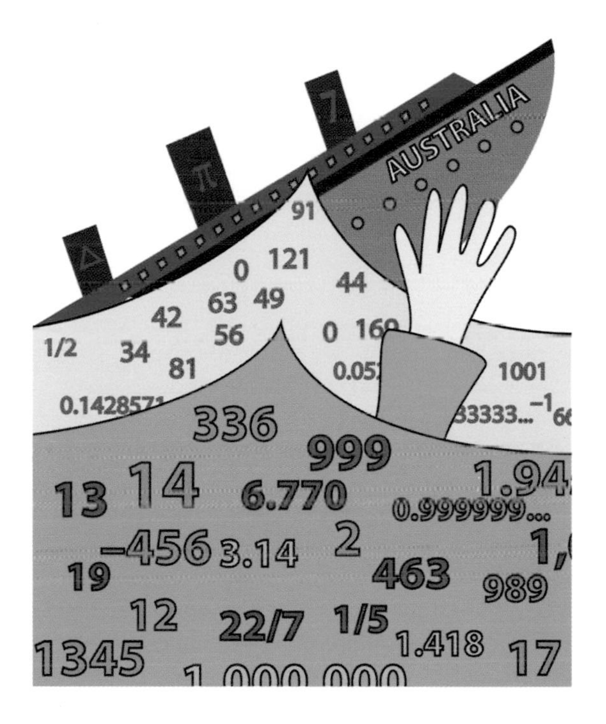

There is a famous and fascinating paradox of movie-making: the French are funny, sex is funny, and comedy is funny, but somehow French sex comedies are never funny.[1] Which brings us to the topic of Australian math education.

It's a depressing topic. It seems that each week brings a new report on Australia's mathematical doldrums. Recently, it's been the "death spiral" of year 12 advanced mathematics: about a 20% drop in student numbers over the previous decade. The decline has been even greater amongst girls, which has come off an already absurdly low base. Mathematical Barbie may still be puzzling away, but she's pretty much on her own.[2]

It's not just a problem of higher year classes. There's hardly a happy buzz over the mathematics offered at any level. Nor is there any happy buzz over what students are learning or, rather, are failing to learn. A careful study of the (almost secret) NAPLAN results reveals that students' arithmetic skills are depressingly weak. International test results are no more encouraging.

---

[1] This joke is due to Matt Groening, from his pre-*Simpsons* days.

[2] See Chapter 56.

But why are things so bad? How can Australian math education fail so dismally to be educational?

The Australian Curriculum can definitely take its share of the blame. One would hope for a strong emphasis upon core mathematics and fundamental skills; just as learning a musical instrument requires practising scales, learning mathematics requires a strong grounding in the basics. Unfortunately, the Australian Curriculum contains the exact opposite emphasis. And any primary school teacher who mimics the Curriculum's denigration of arithmetic skills, who fails to emphasise the fundamental role of times tables and mechanical methods, is condemning their students to life in a mathematical ghetto.

Not that honing skills is all there is to a mathematics education. Though the skills are essential, mathematics must also be presented as a human endeavor, as the search for beautiful, provable truths. Again, the Australian Curriculum misses the point entirely; it exhibits no concern for beauty or history or the human element of mathematics. Instead there are bitsy nothing-topics, aimless investigations, silly pseudo-applications, and an endless stream of tedious statistics.

The senior years of this ridiculous curriculum are just being implemented in Victoria. Some minor improvements have been proposed but the subjects will retain their fundamental flaws, and some grave new flaws will be introduced.

In particular, the statistics that is likely to be incorporated into the senior curriculum is truly appalling. It is pointless for senior students, it's incredibly boring, and it will be a horror to teach. Many students may be fleeing advanced mathematics now, but just watch for the flood of refugees in a couple years.

(Readers who wished to express their views on the curriculum were able to submit comments to the Victorian Curriculum and Assessment Authority (VCAA). But it didn't matter: the VCAA apparatchiks are well practised in listening to no one.)

OK, so the curriculum doesn't make things easy; however, what really matters is what happens in the classroom. Which brings us to the teachers.

Of course no education article is complete without a good round of teacher-bashing. However your Maths Masters will forgo the opportunity. We may have some criticisms, but we also know that most teachers are very dedicated, that they are trying as hard as they can under very trying conditions. And, teachers are already the whipping boys and girls of every deluded government and every pompous education "expert".

Moreover the general standard of mathematics teaching is not the fundamental problem. Consider, for example, the recent news report that 40% of year 7 to year 10 math classes are conducted by a teacher not qualified to teach mathematics. That's a huge problem and definitive proof of policy failure, but the report makes no mention of a related problem that is just as serious.

In fact, there is as much an issue with *qualified* teachers as unqualified teachers. Hints of this problem can be found in an extensive review of Australian mathematics conducted by the Australian Mathematical Sciences Institute (AMSI), and in their accompanying policy document. Unfortunately, the hints are small and well hidden.

The fundamental question is, What does it mean for a mathematics teacher to be qualified?

Of course the answer depends upon the mathematics in question. Teaching a calculus course requires much more background than teaching year 3 arithmetic.

So, some teachers should clearly be expected to have a major in mathematics while for others such a major may be substantially irrelevant. Currently, primary school teachers are "qualified" if they have an education degree, which will typically include a couple of subjects on teaching mathematics, and that's it.

Such subject requirements are simply insufficient to guarantee that a teacher is qualified in any real sense. To teach mathematics well, one must know more than the mathematical topic at hand, the specific techniques to be taught; one must also know *about* the mathematics, why the topic is the way that it is. That involves consideration of the fundamental nature of mathematics and mathematical thought, including a proper appreciation of mathematics' long and difficult history.

Take fractions. They may appear simple, but the arithmetic of fractions is tricky, and tricky to explain. And more than that, it is tricky to even know what a fraction is. It is very difficult to understand that 1/3 does not mean the process of 1 divided by 3, but rather the *number* that results from that process. It is very difficult to realise that 2/6 and 1/3 are *equal*, that they are the exact same thing, and that the weasel word "equivalent", which is almost always employed, has no place here.

It took thousands of years to sort out the nature of fractions, and only afterwards did the arithmetic rules emerge. These rules cannot be properly understood without some appreciation of what fractions really are, and it is an appreciation every mathematics teacher should have. The same is true for almost every concept in school mathematics.

There are few subjects that a potential teacher can take that will reflect and encourage this deeper mathematical sense. True, some Australian mathematics departments offer some excellent, thought-provoking subjects. However many subjects in many mathematics departments are turn-the-handle formulaic silliness, of little benefit to anyone and of absolutely no benefit to a potential teacher. As a consequence, it is very easy to obtain a mathematics major while gaining no sense of what mathematics is about, what it means to think mathematically.

And that's the good news. Mathematics departments may have their problems but, when it comes to mathematics, the education faculties are a disaster.

Education faculties do not teach mathematics. They exhibit no concern that the vast majority of their students lack the mathematical background to teach well. More often than not, the lecturers themselves have a similarly poor mathematical background.

What takes place instead of the investigation of mathematics? Education students learn silly calculator tricks; they wallow in discussion of "numeracy", that absurd, impoverished substitute for mathematics; they are taught prissy rules of mathematical presentation, none of which anyone obeys outside of a classroom; and, they discuss the teaching of mathematics with other people who have no proper sense of what mathematics is.

Do people realise how absurd this all is? Yes, to some extent. Most teachers know they've somehow been cheated; they wish they knew more mathematics, and knew it more deeply. The federal government allocated $12 million for projects designed to improve the training of math and science teachers. And, the AMSI policy document contains the excellent recommendation that the training of primary teachers include two subjects on *mathematics* (as opposed to mathematics teaching) to be taught "in conjunction with" the mathematics department. All this is good.

But, we also feel that people don't really get it, or at least won't admit it. For example, while pointing the finger at education faculties, AMSI makes no mention of the silly cookbook subjects offered by mathematics departments, and how useless they are for prospective teachers. And, AMSI's finger-pointing is much too gentle; the mathematics subjects for teachers should not be taught "in conjunction with" mathematics departments, they should be taught *by* the mathematics departments, with the education faculties nowhere in sight. (AMSI might also have mentioned that plenty of mathematics departments are so wrapped up in presenting cookbook math, they would make a complete hash of teaching mathematics to prospective teachers.) Similarly, all the federal government's projects involve collaboration between education staff and mathematics staff; every project actually concerned with teaching mathematics would be instantly and dramatically improved by the exclusion of the education staff.

The simple truth is, the major stumbling block for mathematics education in Australia is that teachers, qualified or not, don't learn enough mathematics and they don't learn it well enough. Discussion of anything else is pointless until that problem is resolved. And, the second simple truth is that education faculties are so divorced from the study of mathematics, so lacking in understanding and appreciation of mathematics, they can play no meaningful role in fixing the problem except by getting out of the way.

It's all a hilarious state of affairs. Just about as funny as a French sex comedy.

Puzzle to ponder

We all learn that "A negative times a negative is a positive". Why is that true?

Part 13

The critics at work

Though the main target of our grumpy commentary has been Australian math education, at times we took the occasional shot at other nonsense. Of course we felt compelled to attack the golden ratio cult. And it should come as no surprise that Australian journalists are generally no better at covering mathematics than journalists elsewhere. What might be a surprise is the mathematical clumsiness of Australian science. This seems a byproduct of mathematics being viewed in Australia as little more than part of STEM, the grudgingly accepted fourth letter.

CHAPTER 59

The golden ratio must die!

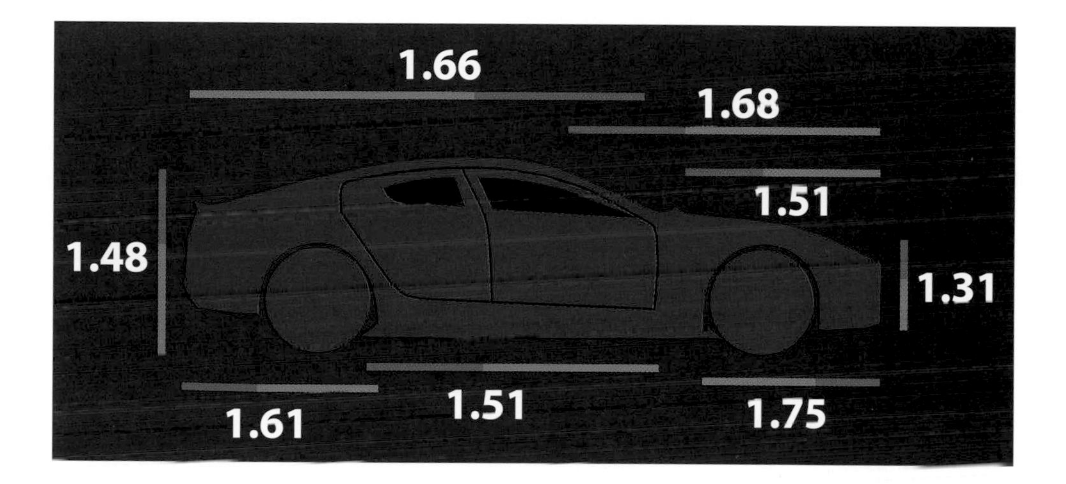

As a child, one of your Maths Masters was fascinated by cars. His favourite was an Aston Martin; the DB5 was a very cool sports car with the added attraction that James Bond drove one. Now, with the release of the Rapide S, these iconic cars are iconic in a special new way: according to *The New York Times*, the latest Aston Martin is designed boot to bonnet in terms of the *golden ratio*.

Sigh! So here we are again.

The golden ratio, about 1.62, is usually denoted by the Greek letter ϕ (phi). It is one of the most famous, and infamous, of numbers. The golden ratio lies at the heart of some beautiful mathematics. Unfortunately, ϕ is also promoted as the number beyond numbers, the key to life, the universe, and everything. It has been the unceasing inspiration for cultish nonsense.

Prior to Aston Martin, it was Venus & Olay applying golden ratio magic to determine stylish dress lengths. (We're fans of Lily Serna,[1] but she, and the supposedly respectable *Australian* newspaper, really should have known better.) And before that, ϕ was heralded as the key to determining the most fertile uteruses. (Alex Bellos and *The Guardian* should have known better as well.[2])

It never seems to end. But now, with the desecration of James Bond's car, we've finally had enough. Once and for all, we're determined to destroy the golden ratio cult.

---

[1]See Chapter 51.

[2]Alex Bellos is a very good mathematics populariser, who writes for the British *Guardian* newspaper.

To begin, let's be clear that ϕ is not simply this magic number that has fallen from the sky, and it is not exactly 1.62. In contrast to the vast majority of media reports, we'll precisely determine our target.

Though the name "golden ratio" appears to have originated in the 19th century, the concept dates back at least to the Greek mathematician Euclid. To facilitate certain geometric constructions, Euclid sought to divide a line segment into a big part and a small part in a special way, as illustrated below.

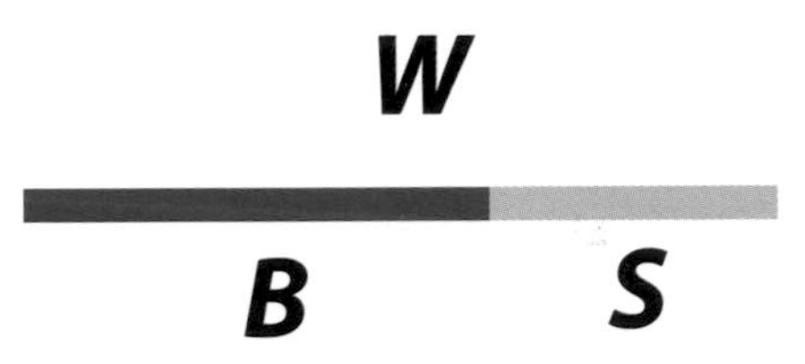

Writing W for the length of the whole line, and B and S for the big and small parts, respectively, Euclid wanted the lengths to be related by the equation

$$\frac{W}{B} = \frac{B}{S}.$$

This common ratio is what we now refer to as ϕ, the golden ratio. A little algebra shows that ϕ is the solution to a quadratic equation, from which it follows that $\phi = (1 + \sqrt{5})/2$, about 1.618, but it is actually an irrational number.

Related to ϕ and with similarly humble Greek origins, is the concept of a *golden rectangle*. A rectangle is said to be golden if it is of just the right shape so that removing a square leaves a (rotated) rectangle of the same shape.

It is not too difficult to show that a rectangle is golden exactly if its length divided by its width is equal to ϕ.

All that is fine, and it leads to some beautiful mathematics. And, the simple ideas underlying ϕ mean that it can elegantly model some naturally occurring phenomena. (Even if other purported appearances of ϕ in nature, such as in the nautilus shell, are utter bollocks.)

But let's talk aesthetics. Is the golden ratio, as is so often claimed, the key to beauty? The answer is emphatically "No!" Moreover, not only is the golden ratio not the key to beauty, it *cannot* be. As we shall explain, the notion is fundamentally absurd.

Of course there are constant claims that great artists and architects have used the golden ratio or golden rectangles in their works. However, almost always these claims are supported by no evidence beyond the exhibition of some ratio that is *roughly* equal to ϕ. And it is not unusual for the ratio to be *very* rough: people hardly ever check.

Consider, for example, the new Aston Martin. The stylish website features the Rapide S with supposedly golden ratios marked every which way. However, as our graphic above illustrates, the segments determining these ratios appear to be quite arbitrary, and many of the resulting ratios are nowhere near ϕ. It may still be true that the Rapide S was designed with the golden ratio in mind, but it is telling that the company's own graphics suggest the opposite.

Now, even supposing some great artist used ϕ in their work, what is the evidence that ϕ is the key to the greatness of their art? What is the evidence that ϕ itself is beautiful? There is none. Not a scrap.

There have probably been hundreds of experiments on people's preferences for certain ratios. Barely a year goes by without some poor university psych class being interrogated over a bunch of rectangles. The main conclusion that can be drawn is that people's stated preferences depend very subtly on how they are asked.

Consider the three rectangles pictured below.

Most people will choose the middle rectangle as the most pleasing. For some that's simply because it is the middle in position; or, it may be because it's the middle in proportion. Or it may well be that there is a general dislike of rectangles that are either too squarish or too skinny. However, even if true, it is an astronomical leap to go from people's vague preference for not-square-and-not-skinny rectangles to precise conclusions about golden rectangles. (In fact, the middle rectangle above has length-to-width ratio of 1.7: it's our first step in starting a new, Seventeen Tenths cult.)

Most experiments that even remotely support ϕ being the key to beauty involve some (probably unconscious) push-polling of the type above. And many experiments offer no such evidence at all. At the end of the "Puzzles to ponder" section is a selection of rectangles from a superb debunking article by mathematician George Markowsky.[3] Have fun choosing your favourite, and have fun finding the golden rectangle.

At this stage it should be clear why any attempt to prove that ϕ is the key to beauty is doomed to failure. An experiment may indicate a commonly held and *rough* preference for certain ratios, and that could be interesting. But an experiment simply *cannot* hone in on a precise number.

The limitation of human senses ensures that no experiment can distinguish a general preference for ϕ over 1.618 or 1.62, or probably even 1.6. There is not a shred of evidence that the golden ratio is any more beautiful than boring old 8/5, and we very much doubt there ever will be.

So, we've done it: the golden ratio is dead and we can all relax. Hardly.

[3] *Misconceptions about the golden ratio*, College Mathematics Journal, 23, January 1992, no. 1, 2–19.

People have been debunking golden ratio nonsense for as long as the nonsense has been around. In 1922, art critic Theodore A. Cook wrote a brilliantly scathing article entitled "A New Disease in Architecture".[4] It said all that was needed about the burgeoning golden ratio cult and the whole absurd notion of reducing beauty to a number: "*There is no short cut to the beautiful, no formula for the creation of the perfect.*"

Cook's article did no good. Markowsky's article did no good. Golden ratio silliness has been debunked over and over and over and over again, and it never does any good. We predict that the next Bond movie will feature James driving his special Aston Martin ϕ-car, and with the beautifully ϕ-proportioned Bond girl in her special ϕ-dress, and... well, we don't dare ponder further.

We'll continue to fight. We'll hammer the ϕ cult whenever it appears. But we are fully aware that it will never do any good. The golden ratio is a zombie number: no matter how many times it is killed, it will always rise again.

Puzzle to ponder

1. Prove the golden ratio is equal to $(1 + \sqrt{5})/2$.
2. Find the golden rectangle in the gallery of rectangles below.

[4] *The Nineteenth Century*, **91**, 1922, pp. 521–532.

CHAPTER 60

Newly done? Bernoulli done!

Johann Bernoulli

A while ago a fantastic math story appeared in the news. It concerned Shouryya Ray, an Indian-born school student living in Germany. Reportedly, Shouryya's entry to a student science competition included the solution to "Newton's 300 year old riddle", an unsolved mathematical problem that had stumped the great Sir Isaac Newton. It was obviously a terrific achievement.

Except, the story was pretty much nonsense.

To be fair, there were credible aspects to the story. It is true, for example, that 300 years ago Newton's best work lay behind him. That's possibly because at that stage he had been dead for 80-odd years. (Usually, the "300 year" headline would later be clarified to an absurdly vague but technically correct "over 300 years".)

Surprisingly, it seems that it was the reputable German newspaper *Die Welt* that started the whole thing. And of course, once a silly story enters Rupert Murdoch's News Limited system, it will be broadcast far and wide. None of the many reports contained a single note of scepticism or made at all clear what Shouryya had supposedly achieved.

The reports included little more than vague references to "projectiles" and "air resistance". It took some clever mathematical detectives in the nerdy corners of the internet to piece together the real story. Later, mathematicians Ralph Chill and Jürgen Voigt were able to review Shouryya's work, after which they produced an excellent summary of what Shouryya had and had not accomplished.

Famously, Sir Isaac Newton clarified the concept of force, in particular gravitational force, and he used this to derive Johannes Kepler's laws of planetary motion. Closer to Earth, if we throw a ball, then it is very accurate to assume that the gravitational force on the ball is everywhere the same. This leads to the familiar, simplified model of projectile motion, with the familiar parabolic paths (first determined by Galileo) as solutions.

Then there's air resistance. Friction from the air will obviously have a retarding effect on a ball's flight, but there is no simple and accurate model of this effect. It depends subtly on the specific properties of the ball and the medium through which it is flying.

Nonetheless, we can still consider simplified models, of which the simplest is obtained by assuming that the air resistance is *linear* with the speed; in this model, doubling the speed of the ball doubles the resistant force against its flight. Including linear air resistance makes for more complicated equations, but they are not too difficult, and Newton was able to solve them.

Unfortunately, the linear model does not tend to be very realistic; Newton himself described the model as "belonging more to mathematics than to nature". Generally more useful is the *quadratic* model of air resistance: in this model, if the speed is doubled, then the air resistance increases by a factor of 4. This is the model Shouryya Ray considered, and this is where things get murky.

No one knows how to solve the equations of projectile motion with quadratic air resistance, at least not in the usual sense that we think of a "solution". Yes, the equations can be thrown into a computer and some numbers will be spat out, and then, yes, pretty graphs can be drawn: we have drawn such a graph below. (It's the red curve, together with comparable graphs of the linear resistance model (blue) and the no-resistance model (green).) However, neither human nor machine has been able to derive what mathematicians call a general *closed form solution*, a simple formulaic solution in the same sense that "parabola" solves the no-resistance problem.

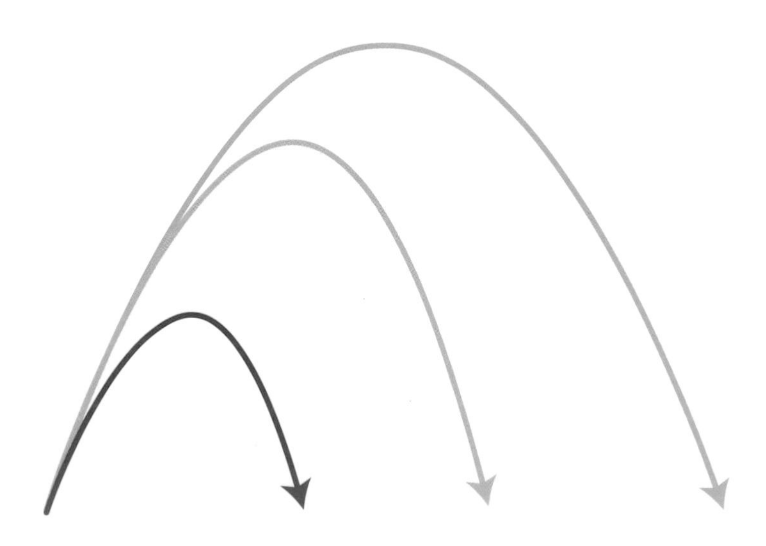

What about Sir Isaac? He argued that the quadratic model was physically natural and he worked on the model, but it is true that he was unable to solve the equations. Moreover, Newton's work included a significant error, which was happily pointed out to him by one of his great rivals, Johann Bernoulli (pictured above).

And Shourrya Ray? As Chill and Voigt make clear, Shouryya "solved" the quadratic resistance problem in a certain sense, but not in the closed form sense, not in the sense that most people would think of as a solution.

Shouryya derived what mathematicians refer to as an *analytic solution*, meaning it includes a particular type of infinite sum (what is known as a power series). Moreover, though Shourrya's work was clever and (mostly) correct, and incredibly impressive for a school student, his results were neither new nor surprising. In Newton's time such work was tricky, but analytic solutions are not in general difficult to obtain, and such a solution to the quadratic resistance problem was essentially derived by Johann Bernoulli in 1719.

So, what should have been an engaging local story about the impressive work of a budding mathematician was transformed into an almighty beat-up. It happened because reporters ignored the warning signs, which were there in neon lights.

The first warning sign was simply the magnitude of the claim. Yes, 300-ish year old problems do get solved,[1] and, yes, young mathematicians sometimes really do spectacular work.[2] But it is rare, much rarer in tandem, and one expects real evidence to back the claim of any such occurrence.

Which points to the major warning sign: there wasn't a mathematician in sight. There was never any indication that expert mathematicians had vouched for the originality or importance of Shouryya's work, and rarely did a reporter bother to check the story with a mathematician. (Early on, ABC radio in Canberra sought your Maths Masters' opinion, and accepted our dampening scepticism in surprisingly good humor.)

This was all a silly episode, but fortunately the reporting of mathematics and science is not always so poor. When it comes to a major issue, such as global warming, news organisations are much more diligent. Then, when it really matters, they appreciate the value of expert opinion, and the importance of distinguishing good science from cultish disbelief. Yeah, sure.

Puzzle to ponder

We are looking for a function $f(x)$ that solves the equation $f(x)(1-x) = 1$. Check that the infinite sum $1 + x + x^2 + x^3 + \cdots$ works. What then does this suggest the value of this infinite sum to be?

[1] Fermat's Last Theorem is perhaps the most famous example.

[2] The famous mathematician George Dantzig accidentally solved two important problems as a first year PhD student when he mistook the writing of the so-far unsolved problems as homework.

CHAPTER 61

Will math kill the rhino?

One of the distressing aspects of the modern world is the large number of animal species in danger of extinction. The authoritative guide to this depressing reality is the *IUCN Red List of Threatened Species*, maintained by the International Union for Conservation of Nature.

Through a careful and extensive process of evaluation, the IUCN categorises highly endangered species as Vulnerable, Endangered, or Critically Endangered. For example, the lion is listed as vulnerable, the tiger as endangered, and the Sumatran rhino as critically endangered.

In 2011 a team of ecologists at the University of Adelaide and James Cook University introduced the SAFE index, a new measure of the threat of extinction. They have suggested that the SAFE index might be used in conjunction with, or even replace, the Red List categories.

Their paper has appeared online in the journal *Frontiers in Ecology and the Environment.* Professor Corey Bradshaw, a team member, has described the SAFE index as "the best predictor yet of the vulnerability of mammal species to extinction".

The SAFE index has received extensive and very favorable coverage. There has been no challenge to Professor Bradshaw's claim that his team has made a "leap forward" in measuring extinction risk. We, however, are unimpressed by the leap.

The SAFE index assigns a number to each species. For example, the lion, tiger and Sumatran rhino have SAFE indices of 0.52, −0.21, and −1.36, respectively.

Lower numbers, particularly negative numbers, are intended to indicate a greater danger of extinction.

What's behind these numbers? In a nutshell, the "leap forward" is the notion that the fewer animals in a species, then the less likely the species will survive. That's it. It's a reasonable idea, at least as a general rule of thumb, though it's hardly rocket science.

True, the SAFE index involves extra calculation. It is estimated that the "minimal viable population" for mammal species is 5000. Then, the population numbers are logarithmically scaled, so that a species population of 5000 has a SAFE index of 0.

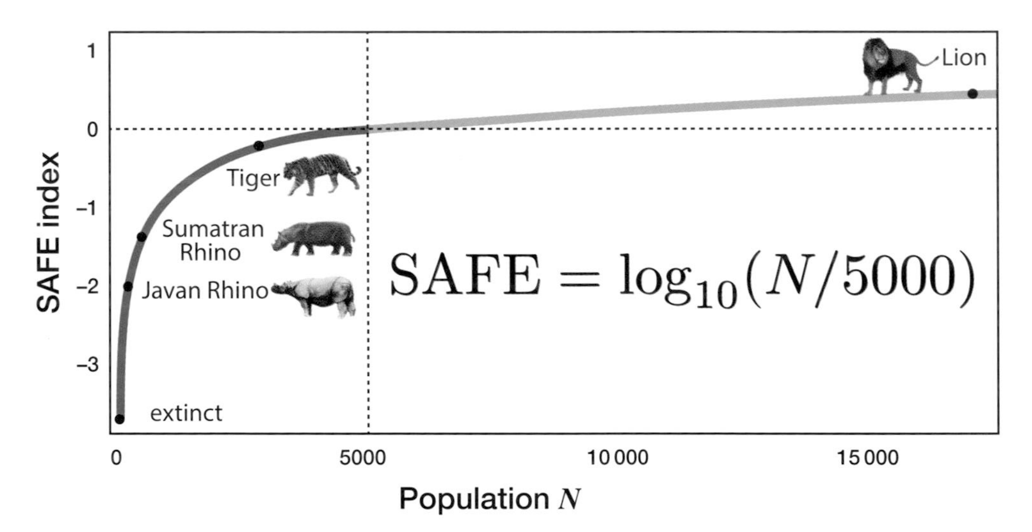

However, we can see no value in this mathematical juggling. The SAFE index contains no more information than the population number of a species, and it is obviously less transparent.

Still, the important question is how the SAFE index (or equivalently, the population number) compares in usefulness to the Red List. The *Frontiers* paper implies that the SAFE index is more objective and is superior by being a quantifiable measure. We are sceptical on both counts.

The focus of the *Frontiers* paper is measuring the agreement of the SAFE index with numerical versions of the Red List. In fact, the Red List itself is the only benchmark of the SAFE index that is indicated: no direct evidence is provided of the ability of the SAFE index to predict survival or extinction.

However, it is impossible to argue that the SAFE index is more reliable, or less subjective, than the standard by which it has been measured. As such, we are aware of no evidence that the SAFE index is superior to the Red List. In fact, we believe that it is inferior, that the SAFE index introduces unwarranted quantification, which is at best distracting and at worst actively misleading.

Consider the *Frontiers* paper's discussion of the Sumatran and Javan rhinos, both critically endangered species. The Javan rhino has a SAFE index of −2.10, lower than the Sumatran rhino's −1.36. On this basis, it is suggested that conservationists may wish to give up on the Javan rhino, to concentrate their efforts upon the Sumatran rhino.

That may be a wise, if tragic, decision. However, it should not be forgotten that what is being discussed here is a total of about 50 Javan rhinos and 250 Sumatran rhinos. It takes no deep thought or logarithmic computation to determine that both species are in dire trouble. Furthermore, for such tiny populations, the precise population numbers lose much of their meaning.

Whatever small chances there are of saving either species of rhino will depend critically upon very local and very specific information, none of which is taken into account by the SAFE index. Indeed, it is difficult to see what the SAFE index contributes here, other than obfuscation.

Puzzle to ponder

What is the SAFE index of humankind?

CHAPTER 62

Shooting logarithmic fish in a barrel

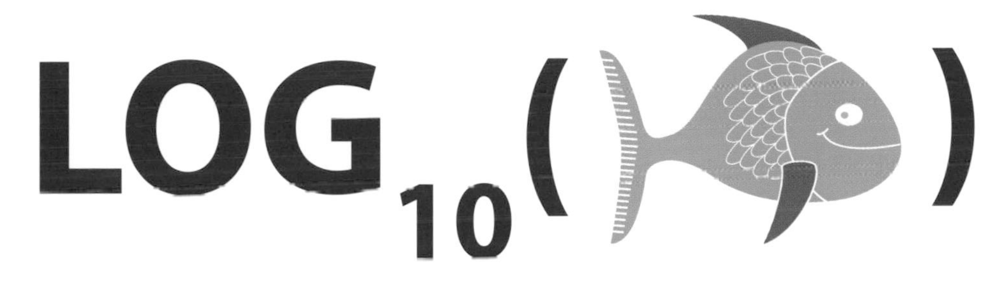

It should be clear by now that our columns vary considerably in style and content: on occasion we have worked hard to try to explain some pretty mathematics, at other times we've been happy to sit back and take easy potshots at mathematical nonsense. Both types of column have been fun and rewarding, at least for us (and we hope for others).

One potshot column that we particularly enjoyed putting together concerned the SAFE index,[1] a naive approach to measuring the vulnerability of endangered species. Though undoubtedly well intentioned, the SAFE index is an archetypal example of mathematical fancy dress camouflaging the thinness of the underlying ideas.

An unexpected benefit of writing the SAFE index column was our meeting Associate Professor Michael McCarthy. Michael is an ecologist at The University of Melbourne, working on conservation biology and ecological modelling, and he clearly grows impatient with some of the fluff that appears in his field. Michael and his colleagues (and others) had submitted their own, detailed critique of the SAFE index. Michael has also performed yeoman's work analyzing the (conservative) Victorian Liberal Party's ground-breaking study into the effect of cows trampling on fragile alpine vegetation.[2]

Michael also alerted us to a new venture, the Ocean Health Index (OHI). True to its name, the OHI is intended to be a measure of the health of the Earth's ocean ecosystem. It was launched in 2012 with an article in the prestigious journal *Nature* and is supported by a large international team of respected marine scientists.

So, the OHI has to be a good thing, right? Perhaps. However, Michael and a number of his colleagues have some pretty large nits to pick. (*Nature* declined to publish their letter in response to the original article, though an independent critique and authors' reply subsequently appeared in the journal.)

---

[1]See the previous chapter.

[2]Reactionary idiot cowboys permitting grazing in fragile terrain, with the paper-thin pretense of "studying" the effects. Australia's version of Japanese "research" on whales.

The OHI attempts to measure the status of ten public goals for a healthy ocean. Included are specifically physical and biological goals, such as water cleanliness and biodiversity, as well as human-centred goals, such as tourism. Each of the ten goals are scored out of 100, and then the scores are averaged to give the overall OHI. The Index is measured for each country as well as for the world overall. For example, Australia's OHI in 2013 was 70 (43rd in the world), compared to the Global Index Score of 65.

What's wrong with that? Plenty, and some of it obvious. Michael is diplomatic, suggesting that the meaning of ocean health is "unclear". Your less diplomatic Maths Masters will be more blunt: the Ocean Health Index is pointless.

To begin, one might conceivably attempt to distill the state of "water cleanliness" or "tourism" to a single number, but then what? If water cleanliness receives a score of 90 and tourism a 10, what can the average of 50 possibly indicate? What if the numbers were reversed?

There is clearly little meaning in the average of ten such numbers. Moreover, the extensive interrelatedness of the goals being scored muddies the little meaning that might exist. It follows that the Ocean Health Index is not a true measure of anything; it is just faking simplicity.

Furthermore, as Michael and others have noted, the ten individual scores are in themselves problematic, and intrinsically so. The OHI has the laudable purpose of measuring ocean health with respect to human interaction. However, the computational effect is that each score is an intermingling of physical and biological measures of the current ocean state with human pressures upon that state. The result is that even the individual scores are scores of nothing in particular, neither fish nor fowl.

It gets sillier. Michael has alerted us to some truly bizarre arithmetic lurking in the Index.

One of the ten goals in the OHI is food provision, being the "the amount of seafood captured or raised in a sustainable way". This is calculated as the weighted average of x_{FIS}, a score of "fisheries" health, and x_{MAR}, a "mariculture" (ocean fishing) score.

The *Nature* article does not indicate how x_{FIS} and x_{MAR} are calculated, however some details are provided in the 2012 supplementary materials. We're simplifying, but x_{FIS} is essentially a ratio, a comparison of the number of fish harvested in a year to the maximum sustainable harvest. To calculate x_{MAR}, one first calculates the number of fish caught divided by the area of the region being fished. (The detailed computation involves the consideration of different fish species and the incorporation of estimates of sustainability.) Then, one takes a logarithm, giving in effect the following formula:

$$x_{MAR} = \log_{10}\left(\frac{\text{Fish}}{\text{Area}} + 1\right)$$

Why the +1? God only knows. (In 2013 the +1 was eliminated from the formula.) Why the logarithm? Again, God only knows. And then, what to make of the overall food provision score: How does one average x_{FIS}, a unitless ratio, with x_{MAR} the logarithm of (1 more than) a fish density? We doubt that even God knows the answer to that one.

Mathematical modelling is difficult and subtle; one should avoid declaring hard and fast rules, and one should be wary of being too, or too quickly, critical. However, it is a fundamental rule to not sum quantities of different physical types. For example, if a car travels 200 meters in 7 seconds, we don't then create a Frankenstein quantity of 207 met-secs. Furthermore, though logarithmic scales can be enlightening, we have never seen logarithms employed in the manner attempted in the Ocean Health Index.

We are sure, despite its glaring flaws, that the Ocean Health Index is well intentioned and incorporates good and important research. It may be salvageable. However, we cannot understand why the creators of the Ocean Health Index chose to include such clunky mathematics in their model. Nor can we understand why *Nature* chose to accept and publish the article in that form.

Whatever the explanation, for us the expression "appeared in *Nature*" no longer has such an authoritative ring.

Puzzle to ponder

We've got a new idea for an index which measures what kind of "shoe person" you are. Our shoe index is the logarithm (base 10) of the average of your shoe size and the number of pairs of shoes you own. One Maths Master has shoe size 8 and owns two pairs of sandals. This gives your Maths Master an average of 5 and a shoe index of approximately 0.7. His wife has a shoe index of 2.3. How many pairs of shoes does she own?

CHAPTER 63

Sliding downwards

It's school holidays and the kids need some activity. What do you do? Well, you can spend an afternoon at Scienceworks, Melbourne's very own interactive science museum.

Recently, some junior Maths Masters did just that. They had great fun, racing wheelchairs, playing a transparent player piano, taking a guided tour of the night sky in the planetarium, pitting their strength against the closing jaws of a T-Rex, and much more. Definitely plenty of entertainment for the whole family.

However, there were aspects that left the more senior Maths Masters much less impressed. Lingering over an exhibit and observing how people actually engage with it, it soon becomes clear that hardly anybody takes the time to actually figure out what is going on.

Visits to Scienceworks clearly consist in the main of pushing any button in sight, watching to see if something flashy happens, and then moving on to look for the next button. Seldom does anybody take the time to read any of the accompanying explanations. So, Scienceworks typically acts as little more than a glorified playground. Still a worthwhile experience, but a great waste of a great opportunity. And yes, there are museums that do it better.

The other, very large source of irritation for your Maths Masters is not difficult to guess: Scienceworks contains almost no exhibits that are even vaguely mathematical. Moreover, for the few semi-mathy exhibits there are, the accompanying explanations fail to mention any of the mathematics and are often misleading or plain wrong.

For example, consider the exhibit pictured below. It consists of three slides starting at a common point and descending to the same level: a spiral slide of length 1.6 meters, a straight slide of length 1.75 meters and a curved slide of length 1.9 meters. A lever allows balls at the top of the slides to be released simultaneously, and the visitor is invited to guess which ball will first reach the bottom.

Here is the accompanying explanation:

> Even though it is longer than the other tracks, the curved slide is the quickest. This is because it is much steeper at the beginning, which means that gravity has a greater effect on the ball causing it to accelerate more quickly and reach its top speed sooner. The spiral track takes the longest even though it is the shortest track, because the sideways movement of the ball down the spiral reduces acceleration and ensures it takes longer to reach its top speed.

The clumsy wording hardly encourages visitors to bother with the explanation, and in this case they'd be wise not to: the explanation is vague, off the point, and wrong.

The comments on the spiral slide are dubious and distracting, but we'll focus upon a more important problem: the issue of "top speed". Consider the following photo of the straight and curved slides, with the red path indicating another possible slide. Would the red slide be (ugh!) "quicker" than the original curved slide?

The very steep drop at the beginning ensures that a ball on the red slide will reach top speed very quickly. So, given Scienceworks' explanation, one may suspect that a ball on the red slide would reach the endpoint B more quickly. That is not the case.

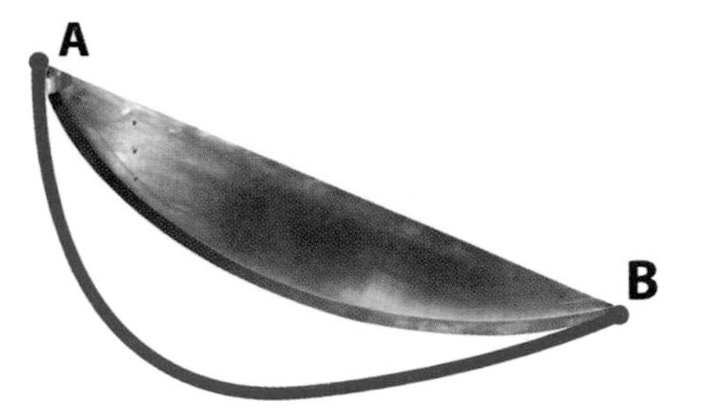

It is hardly surprising that the overall time taken for a trip depends upon much more than the top speed of a trip or upon when that top speed is reached. In fact, contrary to the explanation provided, a ball on Scienceworks' curved slide will reach top speed very late in its journey, at the lowest part of the slide before the little rise at the end.

No thanks to Scienceworks, we happen to know what is actually going on. The motivation for the display is a very famous and important piece of mathematical history, known as the *brachistochrone problem.*

At the end of the 17th century, Johann Bernoulli challenged the leading mathematicians of the day to determine the "curve of fastest descent", the fastest slide connecting two points A and B (assuming perfect sliding under gravity, without friction). The great Sir Isaac Newton accepted the challenge, though he was not amused: "*I do not love to be dunned [pestered] and teased by foreigners about mathematical things...*".

Using the newly invented calculus, Newton and others derived the surprisingly simple answer to Bernoulli's challenge. It turns out that the curve of fastest descent will always be part of a *cycloid*, the curve traced by a point on a the rim of a wheel as it rolls along a straight line.

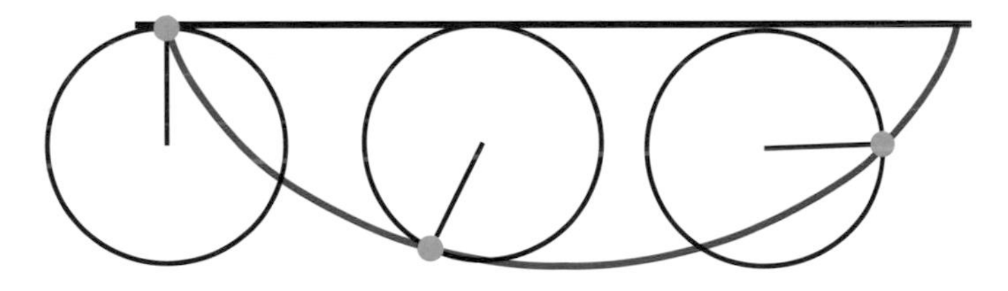

It is no surprise that Sciencework's curved slide fits very nicely into a cycloid. Whoever designed the display knew what they were doing.

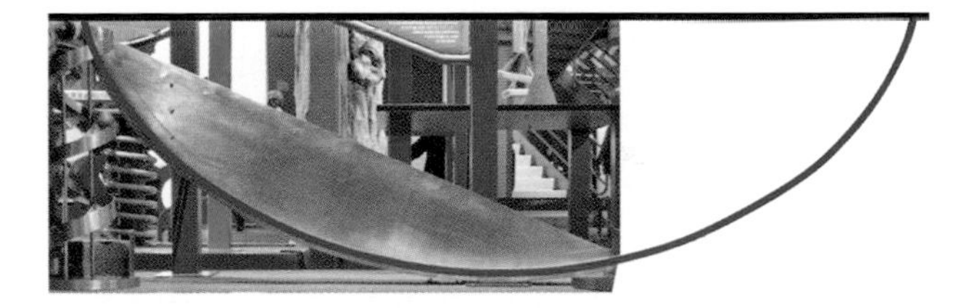

But if the designer knew what they were doing, they apparently didn't think to share this knowledge. Wouldn't it have been preferable for visitors to be offered some of the beautiful mathematics and history, rather than to simply be fed vague, pseudo-mathematical nonsense?

To be fair, this is not a problem specific to Scienceworks. Most science education and science promotion in Australia demonstrates a profound ignorance of, if not contempt for, mathematics. But that's another depressing topic for another depressing day.

Puzzle to ponder

What is wrong with Scienceworks' description of the spiral slide?

CHAPTER 64

Parabolic playtime

In this chapter we want to take a careful look at two of our favourite mathematical toys to demonstrate and to help demystify some parabolic magic.[1]

Many readers will be familiar with our first toy. Whispering dishes are pretty much a must for a science museum, and Melbourne's Scienceworks features an impressive pair. A person whispering a little in front of one of the metal dishes can be distinctly heard by their friend in front of the other dish, many meters away.

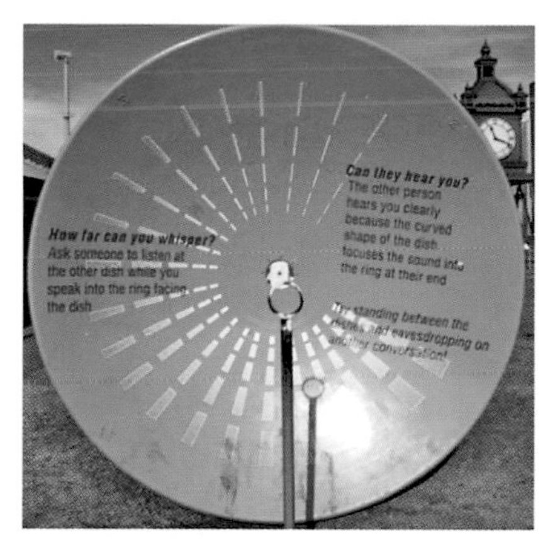

How do the dishes work? According to Scienceworks it's because "the curved shape of the dish focuses the sound". Well, um, thanks.

---

[1]This was our final column. Our editor, evidently eager to pick a fight, secretly altered the wording of our column in a pointless and annoying manner. Our objections were dismissed out of hand, and so we quit. Our very rewarding seven-year relationship with *The Age* newspaper ended because of one clown.

For those who prefer explanations to be a little more explanatory, the dishes are actually paraboloids, the 3D versions of parabolas. Associated to each paraboloid is a *focal point*. As we'll show later, the sound waves emanating from the focal point will bounce off the dish and travel parallel to the dish's axis of symmetry.

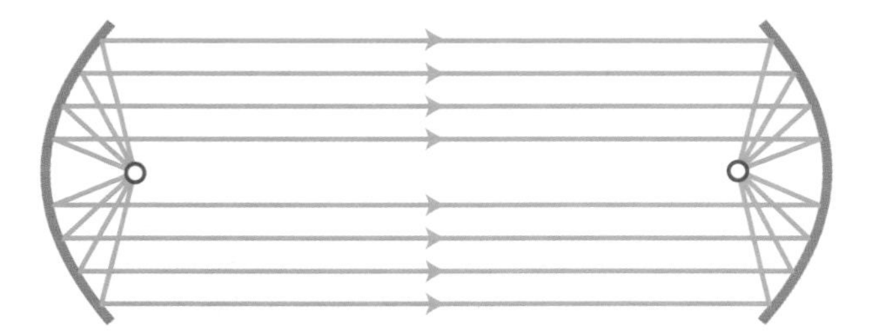

The above diagram indicates how the sound waves will be refocused at the focal point of the second dish. However, something else needs explaining: Why isn't the sound muffled? Specifically, why don't the sound waves leaving the first focal point in different directions then arrive at the second focal point at varying times?

This amounts to asking whether the green paths in the above diagram are all the same length. Indeed they are, though it's hardly obvious, and the question is seldom if ever asked, much less answered. We'll also show this below but let's first play with another toy.

Our second toy comprises two parabolic mirrors, one containing a little hole in the middle.

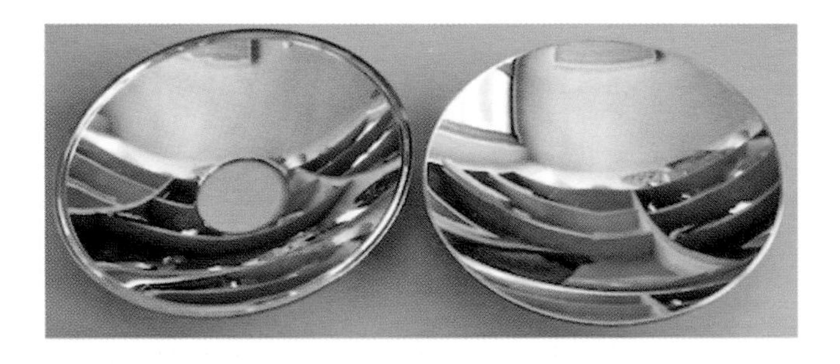

Place a small object like a model car in the centre of the whole mirror, and place the second mirror on top. Then a mirage of the car will appear hovering in the hole.

It's a very impressive illusion and the underlying explanation is pretty much the same as for the whispering dishes:

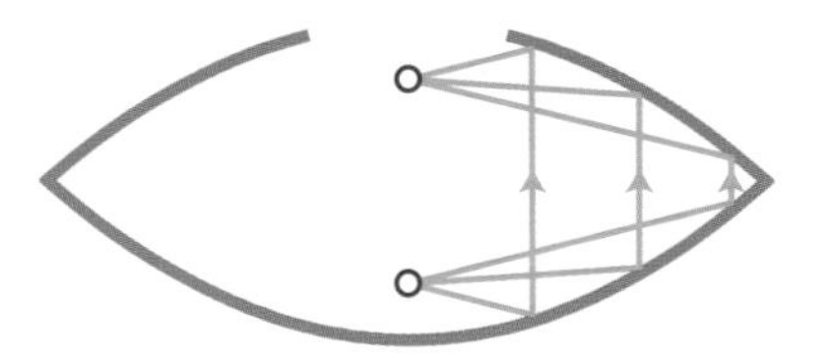

It's all quite amazing and obviously little Maths Masters can have great fun with these toys, even without the "curvy surfaces will focus things" explanation. However your grown up Maths Masters think it's time to get busy, to really figure out how parabolas and paraboloids work. But be warned, dear Reader: the wise masters of the Australian Curriculum have determined that focal points are unworthy of mention, so our discussion will obviously be long or difficult or boring, or all three.[2]

Let's begin with the (hopefully) familiar $y = x^2$ parabola. Associated to this parabola are the red dot and horizontal red line pictured below.

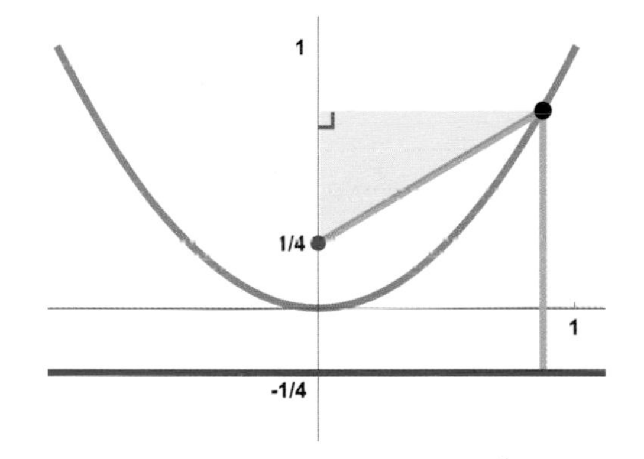

What's special about the dot and the line? A straightforward application of Pythagoras shows that the two brown lines are the same length, which means that any point on the parabola is an equal distance from the dot and the line. This actually defines the red dot as the *focal point* of our parabola, and the red line as its *directrix*. (The ancient Greeks defined a parabola quite differently, as a special slice of a cone. They later proved that any parabola had this focus-directrix property, and our now familiar equation took another 1500 years or so to appear.)

With focal point and directrix in hand, it is now very easy to see the equal-distance property of our whispering dishes. No matter the initial direction, the distance travelled from one focal point to the other is exactly equal to the distance between the two directrices:

Of course this simple argument assumes that the sound waves travel along the indicated paths. To justify this assumption, we have to prove the fundamental property, that a ray leaving the focal point will be reflected and travel parallel to the axis of symmetry. That amounts to showing that in the diagram below the angles A and B are equal.

[2]See Chapters 53 and 58.

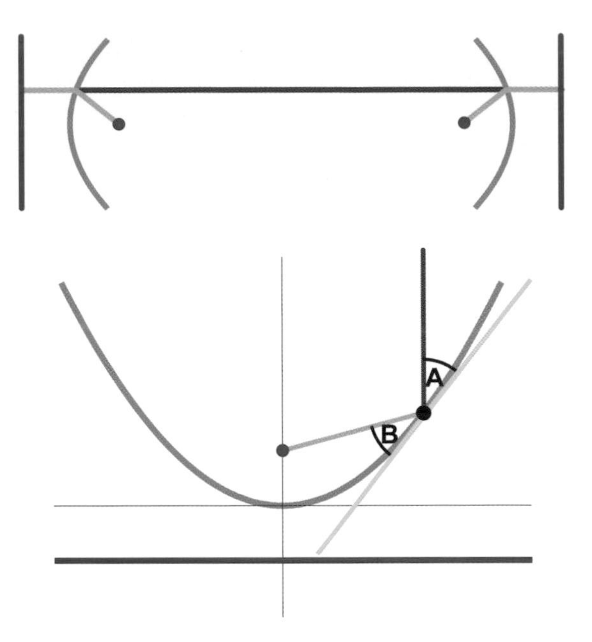

That looks like it may be difficult to prove, but in fact it follows very easily from the diagram below.

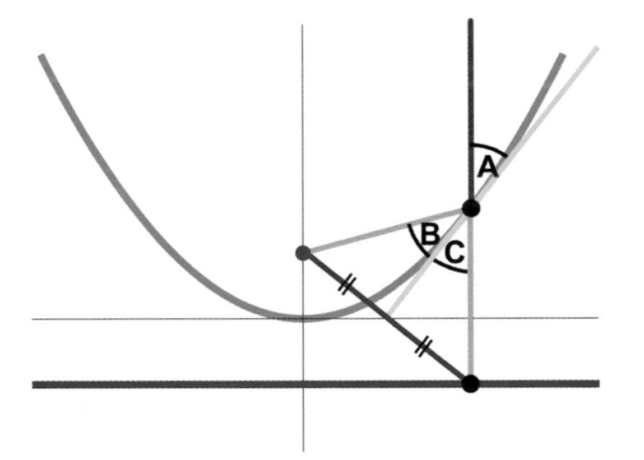

The key observation is that the blue tangent to the parabola bisects the purple segment, as pictured above; this can be proved with one very easy line of calculus. (The calculus can also easily be avoided, by beginning with the purple segment and showing that its perpendicular bisector hits the parabola just once, at exactly the right point.) Now, the two triangles pictured are identical ("congruent"), and so angles B and C are equal. It is even easier to see that angles C and A are equal, and so we can conclude that $A = B$. We're done.

Now, was that long and difficult and boring? It's a mathematical proof, so there will be the well-rehearsed chorus of "yes" from the millions of math-haters. But for anyone receptive to mathematical ideas, the proofs are short, easy, and very beautiful. And they're totally absent from the Australian Curriculum. Go figure.

Puzzle to ponder

Mark a (red) dot on a piece of paper, a couple inches from the bottom edge. Mark a (blue) dot on the bottom edge, then fold and crease the paper through that point, as pictured, so that when folded the bottom edge passes through the red dot. (That is, the pictured red and green dots coincide.) Do this repeatedly, for different locations of the blue dot, making more and more creases. Then the creases will form a parabola. Why?

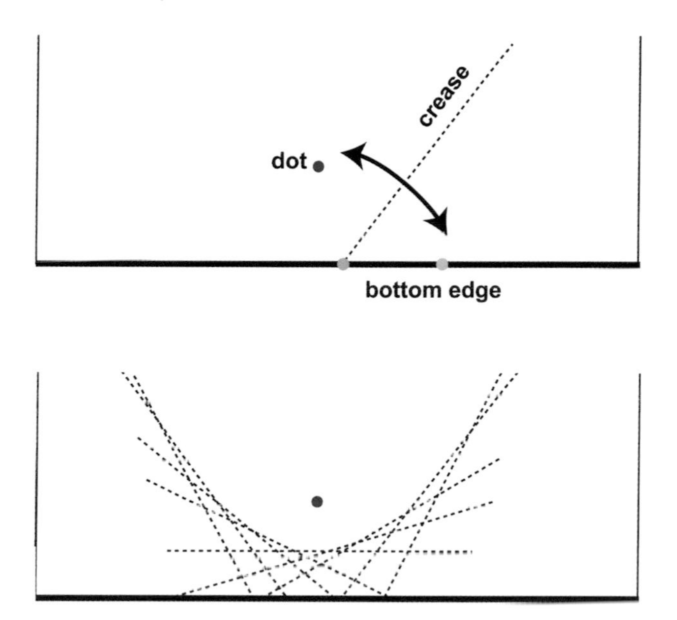

Appendix

Solutions to the puzzles

1. A puzzling Australia Day. All the bits of kangaroos inside the rectangle can be rearranged into eight kangaroos. Therefore, the area of one kangaroo is one eighth of the area of the rectangle, which comes to $13 \times 16/8 = 26$ square centimeters.

2. The biggest pineapple in the world. We are looking for a sequence of numbers in which each number is the sum of the two previous ones, as is true for the Fibonacci sequence. For the Big Pineapple this sequence should end with the three numbers, 13, 13, 26. But that means that the sequence is either 0, 13, 13, 26, or just 13, 13, 26. This means that Woombye's Big Pineapple was never a baby, or that it is itself a very, very large baby.

3. The Nullarbor conundrum. Let's first consider Real Ed. Then the situation is as pictured below.

So, Ed is running towards a point P on the railway line that is a distance a from him and a distance b from the train. The key insight is to realise that Ed wants to choose the point P so that the ratio a/b is as small as possible.

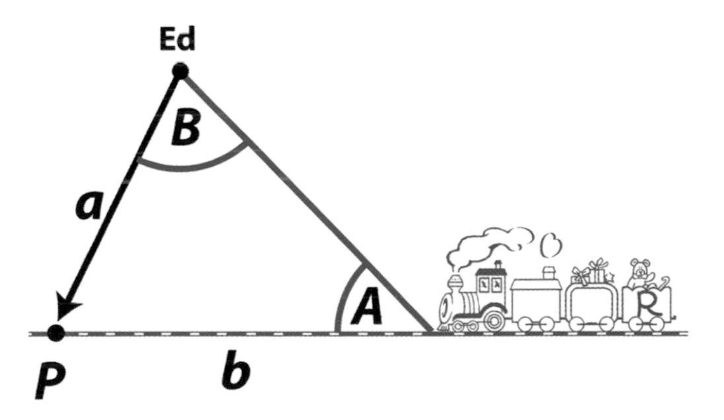

Why? If Ed runs at speed v, then the time taken for him to reach P will be av. Similarly, if the train travels at speed w, then the train will take time bw to get to P. So, Ed will survive as long as $av \leq bw$ or, equivalently, $a/b \leq w/v$. So, no matter the speeds v and w, Ed's best chance is to make a/b small.

But how does Ed know in which direction to go? Notice that the angle A is fixed by the problem, and Ed is choosing the angle B. Then a little trigonometry (the sine rule) shows that to make a/b as small as possible, Ed should ensure that B is a right angle.

However, we can also circumvent the trigonometry. Draw the perpendicular line of length d from P to Q, as pictured below. Notice that, no matter the location

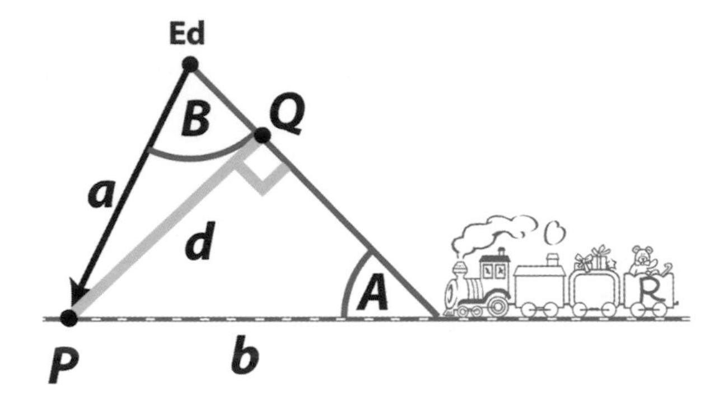

of P along the track, the ratio b/d will always be the same (because the *shape* of the right-angled triangle formed by P, Q and the train is always the same, i.e., the possible triangles are all *similar*). Therefore, since $a/b = (a/d)/(b/d)$, it follows that to minimise a/b, Ed must make the ratio a/d as small as possible, which he obviously does by making B a right angle, in which case $a = d$.

Finally, what about Ideal Ed? The situation is the same, except that now Ed has two, diametrically opposite, perpendicular paths to consider. All he can do is flip a coin, choose one of the two paths and hope for the best.

4. The revenge of the lawn. We've seen that we maximise the area mown if we mow in a straight line. If L is the length of the mowing path, this area would be about $8/10 \times L$ (plus a little extra for the circles at the beginning and end of the path). So, to mow an area of 200 square meters, the path will have to be at least $2000/8 = 250$ meters in length.

5. A Greek in an Italian restaurant. This tiling corresponds to the right-angled triangle having identical smaller sides, and so to Pythagoras's theorem for an isosceles right-angled triangle.

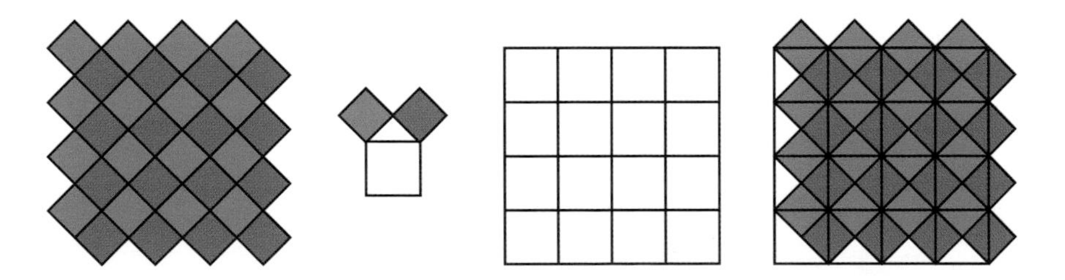

6. Please say hello to Adelaide and Victoria. 1. Below is a Venn diagram that uses four ellipses.

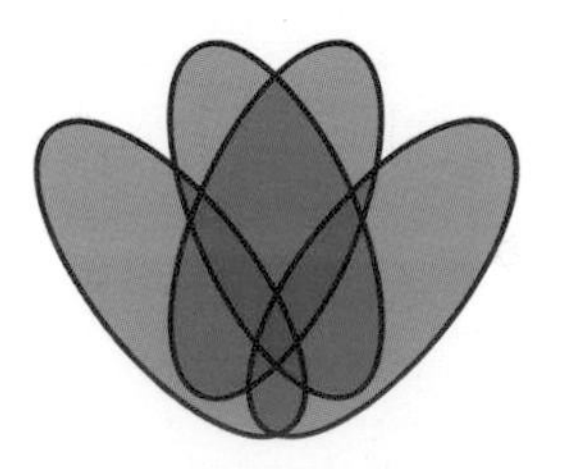

2. The numbers that appear are the binomial coefficients, perhaps most familiarly known as the numbers that appear in Pascal's triangle. For Adelaide and Victoria we'd go to the eighth row of Pascal's triangle, and the numbers are 1, 7, 21, 35, 35, 21, 7, 1.

7. The problem of cricket points. If India hadn't lost a wicket, then they would have 56.6% of their resources left, and so would have used 43.4%. Since 43.4% of Australia's 300 runs amounts to 130.2, India's 150 runs would be superior, making them the winner. If India had lost two wickets, then the calculation would be $300 \times 0.476 = 142.8$ runs, still lower than 150, and so India would still have been declared the winner.

8. Matthew Lloyd vs. Brendan Fevola. Suppose Fevola kicks G goals in 2007. Then his average over the two seasons would be $(G + 84)/40$, and we want that to be higher than Lloyd's average of 75/22. So, the two averages will be the same if $G + 84 = 40 \times 75/22 = 136.3$. So, Fevola needed to kick a minimum of 53 goals.

9. The perfect rugby conversion. The larger the circle through the goalposts, the smaller the (constant) angle to any point on the circle. The circle that just touches the conversion line is the smallest circle that intersects this line. So, the associated angle of this just-touching circle is the largest possible angle from the conversion line.

10. The Socceroos and the group of death. Of course there's no one best answer. However a reasonable approach is to begin by placing the top eight teams into groups A to H, and then the bottom eight teams can go into groups H to A. Then the middle 16 teams can be distributed so that the points in each group are as close as possible to the average of 3732 per group. That still leaves a lot of freedom, but one possible grouping follows.

	top 8	middle 16		bottom 8	total
A	Spain 1485	France 913	Ivory Coast 809	Australia 526	3733
B	Germany 1300	England 1090	Ecuador 791	South Korea 547	3728
C	Brazil 1242	USA 1035	Bosnia 873	Cameroon 558	3708
D	Portugal 1189	Chile 1026	Russia 893	Japan 626	3734
E	Argentina 1175	Greece 1064	Algeria 858	Nigeria 640	3737
F	Switzerland 1149	Belgium 1074	Mexico 882	Iran 641	3746
G	Uruguay 1147	Netherlands 981	Croatia 903	Ghana 704	3735
H	Colombia 1137	Italy 1104	Costa Rica 762	Honduras 731	3734

11. The greatest team of all. If there are just three teams, with no team outclassed, it is easy to see that the three arrows connecting these three teams automatically form a cycle.

With four teams, for no team to be outclassed means that for every team there are either exactly two incoming or two outgoing arrows. Let's focus on a corner with two outgoing arrows as in the following diagram at the top (the same considerations apply to the case of two incoming arrows). Then there are two possibilities for the direction of the green arrow forming the second diagonal of the square. Now the blue arrow is forced, and we can already discern at this instance that there will necessarily be a cycle.

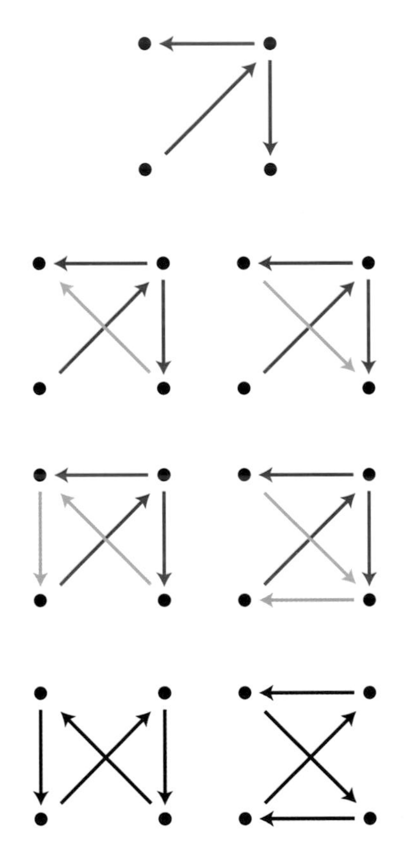

12. Around the world in 80 (plus 129) days. One possibility is to sail from a point to its antipodal point, and then just sail back again along the same route. One rule that would exclude such a silly route would be to require that in addition to including two antipodal points the route also has to include points along all longitudes.

13. A prime mathematician. 1. $4187 = 53 \times 79$.
2. The smallest example is 7, 157, 307, 457, 607, 757, 907.

14. Rubik's cube in ten seconds or less. Assuming an average of 10 seconds solve time and an average of 60 moves per solve, we arrive at a rough estimate of six twists—and one sprained wrist—per second.

15. The wonderful Function of Michael Deakin. Below is a picture of the continental United States coloured with four colours.

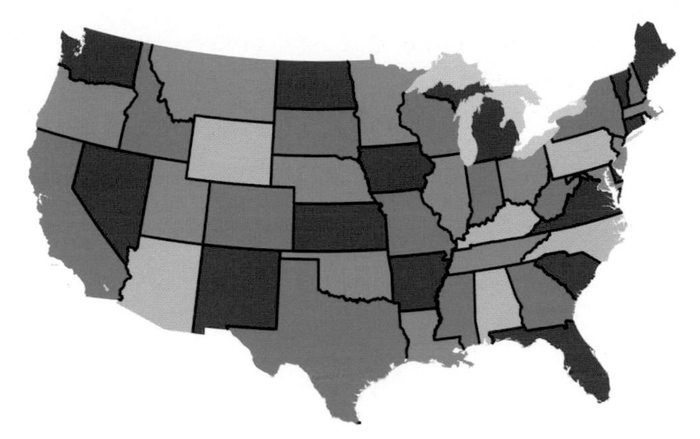

To see that three colours won't suffice, we can focus upon Nevada and its neighbours. We can colour Nevada red, and then we have, say, green and blue to colour the five contiguous states. However, since Nevada has an odd number of neighbours, this is impossible.

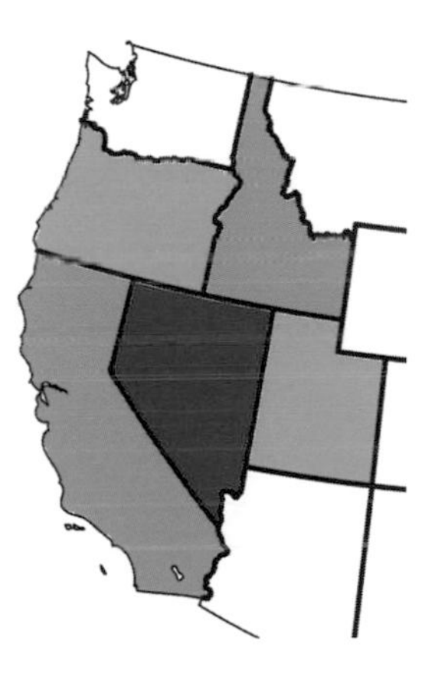

16. A Webb of intrigue. There's eleven in total:

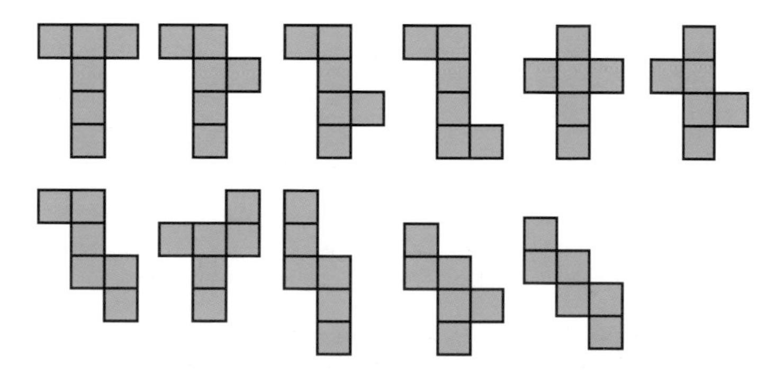

17. A very peculiar Storey. 1. Here is one mistake:

2. The rhombus angles are incorrect: Penrose tiles have smaller angles of 36 degrees and 72 degrees. (We also have no idea what a "continuous surface" is.)

18. Federation forensics. 1. Let's use a 60 cm side as our unit of length. Then the wall has dimensions 30×45, and so has area 1350 square units. As well, each triangle has area $1/2 \times 2 \times 1 = 1$ square unit. So, there must be a total of 1350 triangles.

2. Here are a few rectangles in the patch of the pinwheel tiling under consideration:

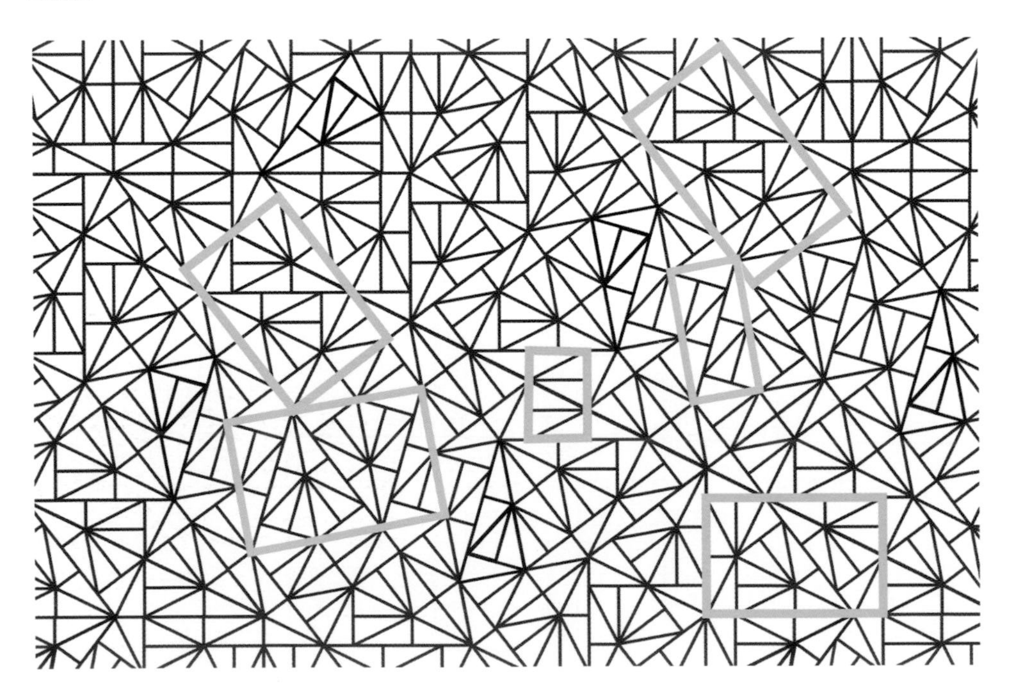

19. More forensicking. 1. An isosceles right-angled triangle is rep-2.

2. Every triangle is rep-4.

3. Finally, suppose that we want a rectangle of dimensions $L \times S$ to be rep-N. We'll determine the proportions so that slicing the rectangle parallel to the short sides gives N rectangles of the same shape. Since the ratio of long to short side is the same for the original rectangle and the sub-rectangles, we find

$$L/S = S/(L/N).$$

Therefore,

$$(L/S)^2 = N.$$

So, a rectangle will be rep-N if the ratio of its sides is $\sqrt{N}$ to 1.

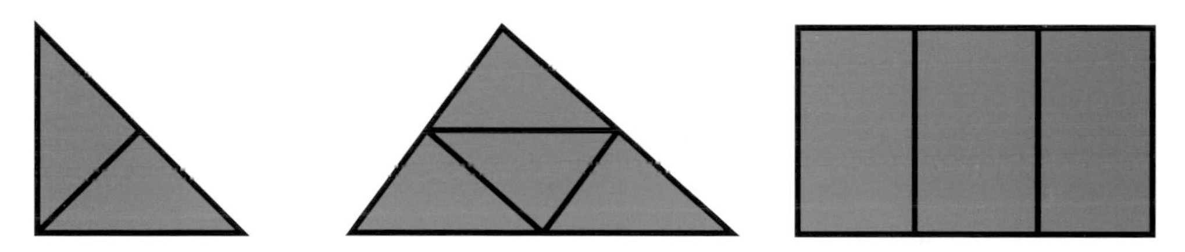

20. Excavating a mathematical museum piece. Here are two examples of Pythagorean quadruples:

$$1^2 + 2^2 + 2^2 = 3^2,$$
$$2^2 + 3^2 + 6^2 = 7^2.$$

21. AcCosted at Healesville. A donut, which mathematicians refer to as a torus.

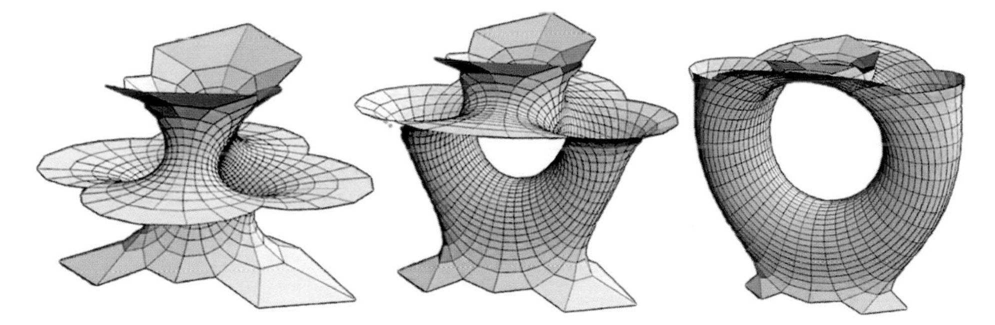

22. The Klein Bottle Beach House. Cut the glass Klein bottle in half along its symmetry plane and the two resulting halves are Möbius strips.

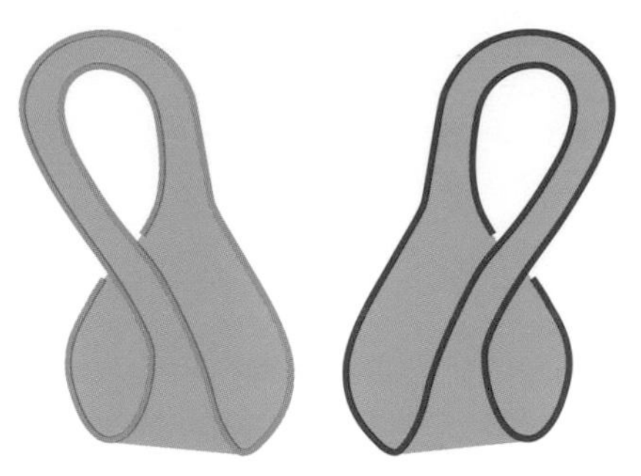

23. Waiting for the apple to fall. Sir Isaac is (almost certainly) doomed to fail. It is a bit of a vague question, so there are some possible loopholes that one could conjure up to help Sir Isaac out (like tossing one apple in the air before getting on the bridge and catching it at the moment you step off the bridge). However, if we are talking about straight juggling, then catching a huge, tossed apple would momentarily exert an extra force on the bridge that would exceed the weight limit.

24. A bucketful of two-up. 1. If there are an odd number of coins in the bucket, this is easy to see; there will either be an even number of Heads and an odd number of Tails, or vice versa, with a 50-50 chance of each. If there's an even number of coins, the easiest thing to do is throw all but one coin. At that stage there's a 50-50 chance of an even or odd number of Heads. Then, throwing the last coin doesn't change these odds.

2. If we begin with $2N$ coins, then the chances of Odds is given by a "binomial probability" calculation, and comes to

$$\frac{(2N)!}{2^{2N}(N!)^2}.$$

Here $N!$ (pronounced "N factorial") stands for the product $1 \times 2 \times 3 \times \cdots \times N$. Now, in theory, we can keep trying different values of N until we wind up with a probability below 1/(8 million). However, in practise, N will have to be huge, and evaluating this formula for huge N is impossible, even with the aid of a good computer. So, we have to approximate. The simplest approach is to use what is known as *Stirling's formula*, a famous and important approximation to $N!$. Using Stirling's formula, it turns out the probability of Odds with $2N$ coins is approximately $1/\sqrt{N\pi}$. (Yes, that's the number π there.) That then gives the number of coins required to be about 20 trillion, which would require one hell of a bucket.

25. We get a Lotto calls. 1. Since $90 = 2 \times 45$, you'd expect that chance to be $1 - (1/e)^2 \approx 0.86$, so 86%.

2. Here is the main step in the "huge" calculation:

$$\frac{1}{1-\frac{1}{\text{huge}}} = \frac{\text{huge}}{\text{huge}-1} = \frac{\text{huge}-1+1}{\text{huge}-1} = 1 + \frac{1}{\text{huge}-1}.$$

This shows that

$$\frac{1}{\left(1-\frac{1}{\text{huge}}\right)^{\text{huge}}} = \left(1+\frac{1}{\text{huge}-1}\right)^{\text{huge}} \approx \left(1+\frac{1}{\text{huge}}\right)^{\text{huge}} \approx e\,,$$

and so,

$$\left(1-\frac{1}{\text{huge}}\right)^{\text{huge}} \approx \frac{1}{e}\,.$$

26. Winning the Lotto by the Melbourne method. The minimum number of System 44 tickets that cover all possible six-number combinations is seven. Here is why. There are a total of 8,145,060 tickets and most of these, namely 7,059,052, are covered by any System 44 ticket. Any System 44 ticket is characterised by the one number it does *not* cover. That is, it covers all six-number combinations except for those that contain this number.

This means that if we buy six different System 44 tickets, we can be sure to have covered all six-digit combinations except for one, consisting of all the numbers excluded by the individual tickets. This means that buying just one more ticket, be it a single ticket or a superticket that contains this combination, will do the job.

27. The super-rigging of poker machines. On any given roll, a total of 7 is most likely, with a 1/6 probability of occurring. However, the probability of *either* a 6 *or* an 8 occurring is $10/36 = 5/18$. Though the details are a little messy (see below), this is the key to the puzzle. Overall, the probability of getting your two 7's first is 3519/7744. A bad bet!

To work through the messy details, we first have to show the following:

(a) The chances of getting a 7 before either a 6 or an 8 is 6/16;
(b) The chances of getting a 7 before a 6 (ditto 8) is 6/11.

(To prove (a), let's call the probability C. Then the first roll of the dice either results in a 7 or (6 or 8) or something else. That implies that $C = \frac{1}{6} + \frac{20}{36}C$, which we can solve to give $C = \frac{6}{36}$.)

Then the chances getting the *two* 7's first is the sum of the chances

$$\begin{aligned}&(7\text{ then }7) + (7\text{ then }(6\text{ or }8)\text{ then }7) + ((6\text{ or }8)\text{ then }7\text{ then }7)\\ &= \left(\frac{6}{16}\times\frac{6}{16}\right) + \left(\frac{6}{16}\times\frac{10}{16}\times\frac{6}{11}\right) + \left(\frac{10}{16}\times\frac{6}{11}\times\frac{6}{11}\right)\\ &= \frac{3519}{7744}\,.\end{aligned}$$

28. How much is a $100 free bet worth? The best option is to bet on a long shot for which the casino's odds are not too unfair. In most casinos that'll be betting on one of the 37 numbers in roulette, which pay off at 35:1. So, 37 coupons will return $35 \times \$5 = \$175$, for an average return of $4.73 per ticket.

29. Tomfoolery and gerrymandering. You'd require at minimum 50 votes in five electorates, totalling to 250 votes. So, a little over a quarter of the votes.

30. Tally up the votes that count. Arrange the preferences for the three voters as in the first table of the chapter:

$$S > P > R, \qquad R > S > P, \qquad P > R > S\,.$$

This amounts to a Rock-Paper-Scissors distribution of the preferences.

31. Green, with envy. 1. No. Whatever Labor does, the Liberal and Greens votes will combine to beat them.

2. 1,001 zombies would suffice. That's enough so that if all zombie preferences went to the Liberals, then the Greens would be eliminated next.

32. Do prime ministers share their birthday cake? 1. Since a group of only 23 people already gives a probability of more than 50%, it is not unreasonable to expect close to certainty in a group of 150, and this is indeed the case. For the exact probability, we simply have to extend the product of fractions in the text down to 216/365. However that's a bit laborious, and it's easy to do a good estimate instead.

Ignoring the first 49 fractions, we'll be multiplying 100 fractions that are at most 316/365. This means that the new product is less than $(316/365)^{100} = 0.00000006$, rounded up. So, the chances of a birthday match will be at least 99.99994%, very close to certainty.

2. Since 435 is greater than 366, the total number of available birthdates, we can be certain that there will members in the U.S. House of Representative who are born on the same day.

3. Finally, prime ministers also share their death days: Francis Forde and James Scullin both died on 28 January. There are a number of presidents sharing death days, including James Monroe, John Adams, and Thomas Jefferson all having died on July 4.

33. The perfect box. It's this simple!

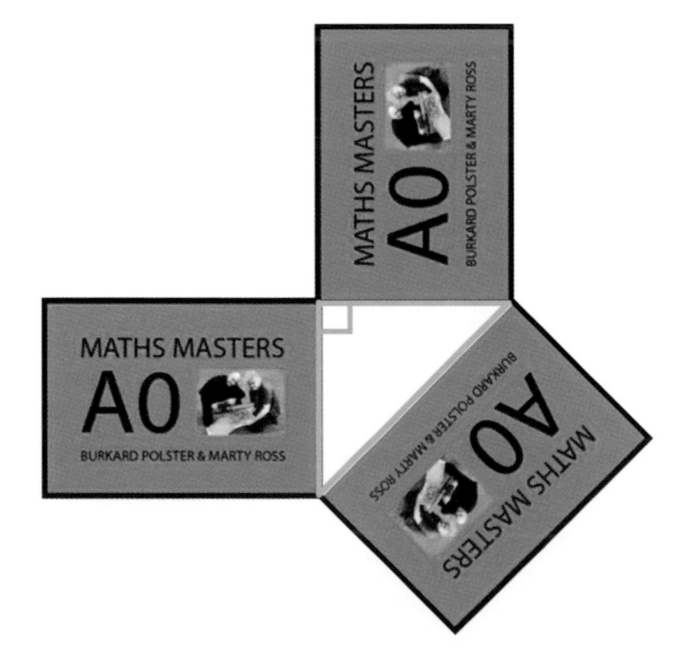

34. Crunching can numbers. If the radius of each can is 1 unit, then the top surface areas are $92+\pi \approx 95.14$ for the rectangular packing and $16+42\sqrt{3}+\pi \approx 91.88$ for the hexagonal packing. So, the hexagonal packing has lower volume, and (just) wins again!

We also note that the hexagonal packing has lower perimeter ($16+8\sqrt{3}+2\pi \approx 36.14$ versus $32+2\pi \approx 38.28$). This implies that, no matter the height of the cans, the total wrapping (including the sides) of the hexagonal packing will be smaller.

35. An added dimension to can crunching. The numbers of circles in the shells are just the first few odd numbers. What this also shows is that the sum of the first N odd numbers is simply N^2.

36. Cool Kepler cat cans. The three triangles above the circle's diameter share this diameter as a common base. So, the triangle with largest area (equal to "half base times height") is the one whose height is greatest. This is the triangle with vertex at the top of the circle, which forms half the square.

37. Taxing numbers take the law into their own hands. Exactly those dollar amounts that start with a 5, 6, 7, 8, or 9 will start with a 1 after conversion. This means that the percentage of numbers starting with 1 after conversion is the sum of the percentages corresponding to 5, 6, 7, 8, and 9: $7.9\% + 6.7\% + 5.8\% + 5.1\% + 4.6\% = 30.1\%$. And lo and behold, this is just the percentage corresponding to 1 as predicted by Benford's Law!

38. Right on the money for a change. No such attempt by the customer to game the system will have any effect on the overall outcome. This is because for every single possible cent value, the average outcome of a gamble is exactly that particular cent value. For example, for 56 cents the probability for rounding up to \$1 is 56/100 and the probability for rounding down to \$0 is 44/100. This means that the average outcome of the gamble is $\$1 \times 56/100 + \$0 \times 44/100 = 56$ cents.

39. Triangulating the Queen. It's not hard to guess that the circle has greatest area, though it's not so easy to prove. It turns out that all shapes of a given constant width have the same perimeter length. But then, among all shapes of a given perimeter, the circle has the largest area.

40. Sex, lies, and mathematics. If F stands for the number of friendly people and N for everybody else, then

$$F + N = 10,000.$$

Two heads come up with a probability of 1/4. So about 3/4 of the friendly people and about 1/4 of the other people will say "Yes". So, approximately,

$$\frac{3}{4}F + \frac{1}{4}N = 6000.$$

We now solve these two *simultaneous equations*. Multiplying the second equation by 4 and subtracting the first equation, we find $2F = 14,000$. So, $F = 7000$, and $N = 3000$.

41. Mathematical matchmaking. Using the recipe in the text, we'll try to make the girls as happy as possible, and we end up with one possible pairing: Angelina–Brad, Julia–George, Nicole–Johnny. It is easy to check that any person who did not get their favourite choice won't have any luck trying to instigate a swap. It also turns out that this is the only stable matching, and in particular beginning with trying to make the boys happy leads to the same pairings.

42. Living in the zone. Here is the Voronoi diagram of the "hexagonal" arrangement of schools:

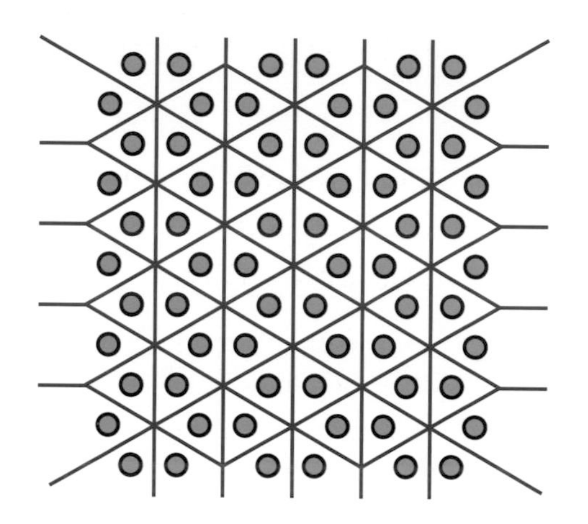

43. Proving a point on speed. Infinity kilometers per hour. To average 100 kilometers per hour (or faster) overall, he would have to pass through the third gate no later than 2 pm. But we know that he's only at the second gate at that time.

44. The Maths Masters' Tour de Victoria. Take the squares of our chessboard to have sides of length 1. We now begin with two observations:

(1) The shortest possible connection between any two cities are vertical and horizontal line segments of length 1;
(2) A round trip will contain exactly 64 straight line segments.

This means that if we can find a round trip made up of 64 line segments of length 1, then we can be sure that it's of the least possible length. There are many such round trips, for example the route pictured below:

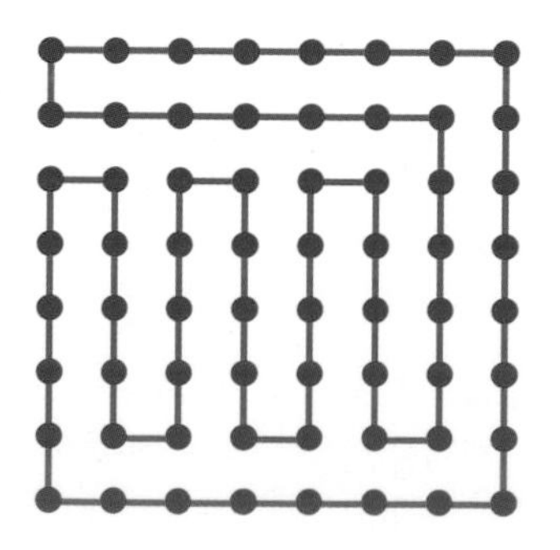

45. The Ausland paradox. Let D stand for the distance between the two cities. The trucks then meet at a point a distance $D/2 - 12$ from Melbaville and a distance $D/2 + 12$ from Sydtown, and they take the same time to get there. If we now let M be the speed of the truck leaving Melbaville and let S be the speed of the truck leaving Sydtown, it follows (by rearranging "speed = distance/time"), that

$$(D/2 - 12)/M = (D/2 + 12)/S.$$

Then, the Sydtown truck (with the Melbaville driver) takes 9 hours to travel the remaining distance $D/2 - 12$ to Melbaville. So,

$$S = (D/2 - 12)/9.$$

Similarly,

$$M = (D/2 + 12)/16.$$

We can then eliminate S and M from the first equation, leaving an equation for D alone. After a little fun algebra (cross multiplying, difference of two squares), we find that D equals 168 kilometers.

46. Schnell! Snell! The shortest path consists of two straight line segments that make the same angle with the water line. Similar to the lifesaver problem, this minimal solution is the path that a ray of light would take, if reflected off of a mirror positioned along the water's edge.

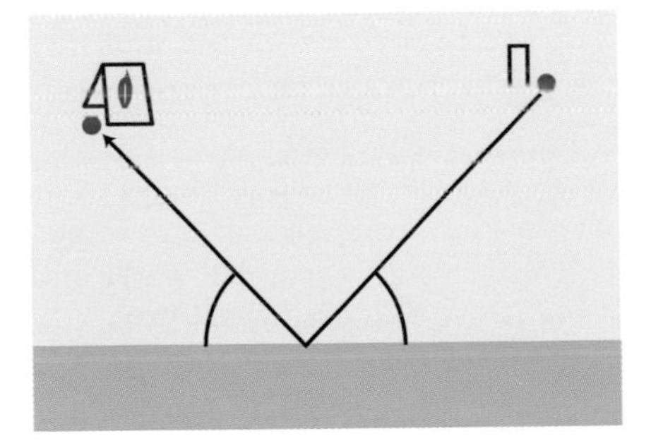

47. The Choice is Right! The contestant should always choose an extreme value. For example, if the value has been narrowed down to be between \$25,723 and \$25,792, the contestant should then choose either \$25,724 or \$25,791. The point is, assuming no precise knowledge of the prices, every value is equally likely. However, choosing an extreme value narrows the choice for the other contestant as little as possible. Playing this way against a player choosing randomly results in the smart player winning about 2/3 of the time. Of course if both players are smart, the contest is again a (roughly) 50-50 game, which will last on average for 51 guesses. Riveting television!

48. Maths MasterChefs. You can think of the picture as a large blue rectangle, parts of which have been covered by two right-angled triangles. Moving the two triangles together, as pictured below on the right, changes the shape of the uncovered blue area but not its area. Therefore the blue parallelogram on the left has the same area as the rectangle with the same base, on the right.

49. Maths Masters under the Scope. Pour 1/3 liter of the cordial into the water, mix and then pour 1/3 liter of the mix back into the remaining cordial.

50. How to murder a mathematician. Euler's identity, $e^{i\pi}+1=0$, is very prominent, and just visible in the lower left corner is the famous identity

$$1+2+3+\cdots+N=\frac{N(N+1)}{2}.$$

51. A show devoted entirely to numbers (and letters).

1. $75-50+4=29$, $75+50+4=129$, $(75-50)\times 9+4=229$, $50\times 8-75+4=329$, $(50+9+4)\times 8-75=429$, $50\times 9+75+4=529$, $75\times 9-50+4=629$, $75\times 9+50+4=729$, $(75+8+4)\times 9+50-4=829$, $(9+8)\times 50+75+4=929$.

2. All the numbers from 1 to 100 can be made up from $1, 3, 9, 27, 81$ using only addition and subtraction. In fact, every one of our target numbers can be represented in exactly one such way. More generally, every integer can be represented in exactly one way with the powers of 3 using addition and subtraction. This fact is at the basis of the so-called *balanced ternary number system.*

52. Bloody Numbers! 1 centimeter for the orange segment and 1.4 centimeter for the blue one means that the angle A in our discussion is about 45 degrees. This translates into the following picture, from which it follows that the wound occurred about 1.5 meters above the spot, right in the middle of the two blood splatters.

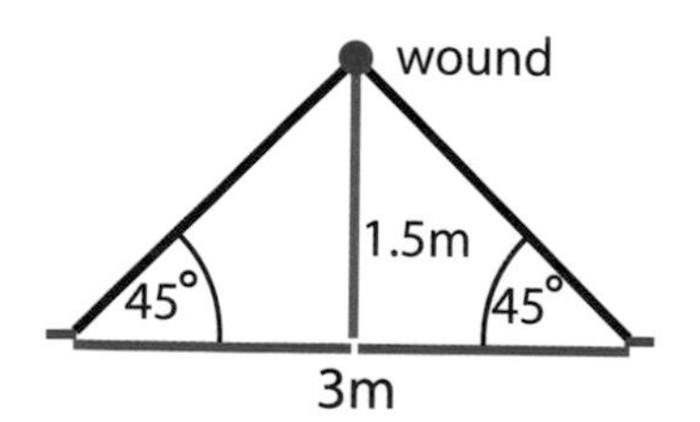

53. Irrational thoughts. This just amounts to the square of a fraction being another fraction. Supposing C is a not a whole number, then we can write $C = m/n$ with $n > 1$, and m and n have no common factors. We'll show this turns out to be impossible. ("Proof by contradiction".)

To do this, we can multiply both sides of the Pythagorean equation by n^2, giving

$$n^2A^2 + n^2B^2 = m^2 .$$

But now it's easy to see this equation is impossible. Remembering that n is supposed to be at least 2, we can look at any prime factor p of n. Then p is a factor of the left side of the equation. But p can't be a factor of the right side, because m and n have no common factors. Contradiction.

54. The best laid NAPLAN. 1. Going through the prime numbers, we find that 7 is a factor of 2009, and then it easily follows that $2009 - 7^2 \times 41$. (Note that a calculator was permitted. Sigh.) However, 2011 is prime, which is definitely not easy to figure out.

2. a) Who cares? b) God only knows. c) God only knows.

55. Divide and conquer. 1. Who cares?

2. About 50% less than one would wish.

56. Educational Barbs. Yes. Since $2/6 = 1/3$, since they're absolutely the exact same thing, it follows that anything you do to $2/6$ is equal to the same thing done to $1/3$. On the other hand if you pretend this question is about index "laws" (which it is not, since they don't apply here), it is easy to "conclude" the equation is false. That's what happened with a prominent Victorian school text, sheepishly followed by thousands of Victorian teachers and students.

57. The statistical problem of greedy pigs. Never. Suppose you've had N rolls so far, and rolled a 6 every time. Then you'll have the maximum possible total at that stage, equal to

$$6 + (2 \times 6) + (4 \times 6) + \cdots + \left(2^{(N-1)} \times 6\right) = (2^N - 1) \times 6 .$$

That's the most you can be risking on the next $(N + 1)$-st roll, with a $1/6$ chance of losing it all. However, the average increase if you avoid the killer 2 is

$$2^N(1 + 3 + 4 + 5 + 6)/6.$$

So, on average your overall gain from the extra roll is at least

$$2^N(19/6) - 2^N + 1 ,$$

which is positive. So, you should always risk another roll of the die.

But of course "never" can't really be the right answer, can it? Figuring out what's going on here, and what's wrong here, warrants its own separate book.

58. The paradox of Australian mathematics education. It's not all that easy to see.

There are two general approaches to making sense of negative numbers (and "new numbers" more generally). One approach is to somehow make ("construct") negative numbers and the arithmetic operations on them, and then check that the normal algebraic rules apply; in this case we would simply *define* the product of two negatives to be a positive. A second ("axiomatic") approach, more natural in the school context, is to just assume that the negative numbers and 0 are there, and that the normal algebraic rules apply.

With this second approach we can prove, for example, $-1 \times -1 = 1$ as follows. To begin, we can prove that (and/or just accept that)

$$0 \times -1 = 0$$

and

$$1 \times -1 = -1\,.$$

Next, we can calculate

$$(-1 \times -1) + (1 \times -1) = (-1 + 1) \times -1 = 0 \times -1 = 0\,.$$

So,

$$(-1 \times -1) + (-1) = 0\,,$$

and adding 1 to both sides gives $-1 \times -1 = 1$.

59. The golden ratio must die! 1. Beginning with the equation $W/B = B/S$ and substituting $W = B+S$, we have $1+S/B = B/S$. The definition $\phi = B/S$ then gives $1 + 1/\phi = \phi$. Multiplying both sides by ϕ, we arrive at the quadratic equation $1 + \phi = \phi^2$. Solving, and ignoring the negative solution, we obtain our rooty expression for ϕ.

2. Fourth row, fifth column.

60. Newly done? Bernoulli done! Just do it!

$$(1-x)(1+x+x^2+x^3+\cdots) = (1+x+x^2+x^3+\cdots) - (x+x^2+x^3+\cdots) = 1.$$

However, just dividing the original equation by $1-x$ gives $f(x) = 1/(1-x)$. This suggests that

$$1/(1-x) = 1 + x + x^2 + x^3 + \cdots\,,$$

which is indeed correct if x is between -1 and 1.

61. Will math kill the rhino? A quick Google search puts the number of humans on the planet at around 7.5 billion. This means that our SAFE index is about 6.2 which means that nothing can happen to us. Right? Right?

62. Shooting logarithmic fish in a barrel. Shoe sizes change from country to country, but of course that does not invalidate our shoe index, and we are confident that *Nature* will eagerly publish our groundbreaking measure of people's shoeiness. Anyway, if Mrs. Maths Master has P pairs of shoes and shoe size S, then

$$\log_{10}((P+S)/2) = 2.3.$$

Unravelling the logarithm, we find that

$$P + S = 2 \times 10^{2.3}.$$

In Australia, women's typical shoe sizes range from 3 to 10. Assuming that Mrs. Maths Masters has a typical shoe size, we conclude that we are talking about somewhere between 190 and 197 pairs of shoes. Amazingly, some critics will undoubtedly continue to claim that our shoe index contains no valuable information.

63. Sliding downwards. Everything.

64. Parabolic playtime. Each crease is the perpendicular bisector of the line joining the green and red dots. But then the last diagram of the chapter indicates that the red dot is the focal point of a parabola, with the bottom edge of the paper being the directrix. And, the creases are then exactly the tangents to this parabola.

Selected Published Titles in This Series

106 **Burkard Polster and Marty Ross,** A Dingo Ate My Math Book, 2017
103 **Ted Hill,** Pushing Limits, 2017
102 **Roger Cooke,** It's About Time, 2017
101 **Anna R. Karlin and Yuval Peres,** Game Theory, Alive, 2017
100 **Patricia Hersh, Thomas Lam, Pavlo Pylyavskyy, and Victor Reiner, Editors,** The Mathematical Legacy of Richard P. Stanley, 2016
99 **Hung-Hsi Wu,** Teaching School Mathematics: Algebra, 2016
98 **Hung-Hsi Wu,** Teaching School Mathematics: Pre-Algebra, 2016
97 **Richard Evan Schwartz,** Gallery of the Infinite, 2016
96 **Erica Flapan,** Knots, Molecules, and the Universe, 2016
95 **Koryo Miura, Toshikazu Kawasaki, Tomohiro Tachi, Ryuhei Uehara, Robert J. Lang, and Patsy Wang-Iverson, Editors,** Origami^6, 2015
94 **Philip J. Davis,** Unity and Disunity and Other Mathematical Essays, 2015
93 **Linda Keen, Irwin Kra, and Rubí E. Rodríguez, Editors,** Lipman Bers, a Life in Mathematics, 2015
92 **Olli Lehto,** Lars Ahlfors — At the Summit of Mathematics, 2015
91 **Jennifer Granville, Sarah Jones Nelson, AnnMarie Perl, Miroslav Petříček, Klaus Friedrich Roth, and Michael Spencer,** Art in the Life of Mathematicians, 2015
90 **Richard Evan Schwartz,** You Can Count on Monsters, 2015
89 **Steven G. Krantz,** How to Teach Mathematics, Third Edition, 2015
88 **Reuben Hersh,** Peter Lax, Mathematician, 2015
87 **Gregory V. Bard,** Sage for Undergraduates, 2015
86 **Boris A. Khesin and Serge L. Tabachnikov, Editors,** ARNOLD: Swimming Against the Tide, 2014
85 **V. I. Arnold,** Mathematical Understanding of Nature, 2014
84 **Richard Evan Schwartz,** Really Big Numbers, 2014
83 **Reuben Hersh,** Experiencing Mathematics, 2014
82 **Lawrence C. Evans,** An Introduction to Stochastic Differential Equations, 2013
81 **Terence Tao,** Compactness and Contradiction, 2013
80 **Elaine McKinnon Riehm and Frances Hoffman,** Turbulent Times in Mathematics, 2011
79 **Hung-Hsi Wu,** Understanding Numbers in Elementary School Mathematics, 2011
78 **Jeffrey C. Lagarias, Editor,** The Ultimate Challenge, 2010
77 **Terence Tao,** An Epsilon of Room, II, 2011
76 **Barbara Gellai,** The Intrinsic Nature of Things, 2010
75 **John B. Conway,** Mathematical Connections, 2010
74 Modelling in Healthcare, 2010
73 **George G. Szpiro,** A Mathematical Medley, 2010
72 **Martin Aigner and Ehrhard Behrends, Editors,** Mathematics Everywhere, 2010
71 **Alexandre V. Borovik,** Mathematics under the Microscope, 2010
70 **Mark Saul,** Hadamard's Plane Geometry, 2010
69 **Herbert Edelsbrunner and John L. Harer,** Computational Topology, 2010
68 **Paul Pollack,** Not Always Buried Deep, 2009
67 **Terence Tao,** Poincaré's Legacies, Part II, 2009
66 **Terence Tao,** Poincaré's Legacies, Part I, 2009

For a complete list of titles in this series, visit the
AMS Bookstore at **www.ams.org/bookstore/mbkseries/**.